CNO ISOTOPES IN ASTROPHYSICS

ASTROPHYSICS AND
SPACE SCIENCE LIBRARY

A SERIES OF BOOKS ON THE RECENT DEVELOPMENTS
OF SPACE SCIENCE AND OF GENERAL GEOPHYSICS AND ASTROPHYSICS
PUBLISHED IN CONNECTION WITH THE JOURNAL
SPACE SCIENCE REVIEWS

VOLUME 67
PROCEEDINGS

CNO ISOTOPES IN ASTROPHYSICS

PROCEEDINGS OF A SPECIAL IAU SESSION HELD ON
AUGUST 30, 1976, IN GRENOBLE, FRANCE

Edited by

JEAN AUDOUZE
*Laboratoire René Bernas du Centre de Spectrométrie Nucléaire et de
Spectrométrie de Masse, 91406 Orsay, France
and Observatoire de Meudon, 92190 Meudon, France*

D. REIDEL PUBLISHING COMPANY

DORDRECHT-HOLLAND / BOSTON-U.S.A.

Library of Congress Cataloging in Publication Data

Main entry under title:

CNO isotopes in astrophysics.

    (Astrophysics and space science library ; v. 67)
    Bibliography: p.
    Includes indexes.
        1.   Carbon—Isotopes—Congresses.   2.   Nitrogen—Isotopes—Con-
gresses.   3.   Oxygen—Isotopes—Congresses.   4.   Nuclear astrophysics—
Congresses. I.   Audouze, Jean.   II.   International Astronomical Union.
III.   Series.
QB463.C18   523.01'9   77—1234
ISBN-13: 978-94-010-1234-8   e-ISBN-13: 978-94-010-1232-4
DOI: 10.1007/978-94-010-1232-4

Published by D. Reidel Publishing Company,
P.O. Box 17, Dordrecht, Holland

Sold and distributed in the U.S.A., Canada and Mexico
by D. Reidel Publishing Company, Inc.
Lincoln Building, 160 Old Derby Street, Hingham,
Mass. 02043, U.S.A.

TABLE OF CONTENTS

PART VI - CONCLUSION

PREFACE

On behalf of the IAU Commission 48 "High Energy Astrophysics" a special session on "CNO Isotopes in Astrophysics" was held at the General Assembly of IAU in Grenoble on August 30, 1976. This topic was chosen since it has recently included many exciting developments in various domains of physics and astrophysics such as for instance meteoritical studies, radioastronomy measurements of the interstellar gas, new determinations of nuclear reactions cross-sections and fractionation processes. The papers of this volume are the written versions of the talks which were presented during this session and cover the most recent observations and theoretical developments regarding this fast-growing field.

Let me thank first Professor Martin Rees (Institute of Astronomy at Cambridge) the chairman of IAU Commission 48, who has sponsored this special session and made it possible. I am indebted to the authors of the papers for their contributions and all the work they have put in the excellent presentation of the material. I owe special thanks to Professor William A. Fowler (Kellogg Laboratory at Caltech) who has superbly chaired this session and made it very exciting and lively, to Professor Beatrice M. Tinsley (Yale University) who accepted the difficult task to conclude and summarize on such a burning and moving topic, and to Dr James Lequeux for his invaluable help in the edition of several contributions.

Finally, the edition of this book would not have been possible without the skilled and generous aid of Miss Jeannette Caro.

Let me express to all of them my gratitude.

Orsay and Meudon   Jean Audouze
November 1976

LIST OF CONTRIBUTORS

(The speakers at the special Conference held on August 30, 1976 at the IAU Assembly – Grenoble, France – are marked with an asterisk)

Audouze J.   Laboratoire René Bernas, B.P. 1, 91406 Orsay, and
                      Observatoire de Meudon, 92190 Meudon, France.

Auer L.H.   High Altitude Observatory, Box 3000, Boulder,
                      CO 80303, U.S.A.

Bieging J.H.   Max Planck Institute für Radio Astronomie Auf dem
                      Hugel 69, 53 Bonn 1, West Germany.

Blanco A.   Instituto di Fisica, Universita di Lecce, Lecce,
                      Italy.

Bussoletti E.   Instituto di Fisica, Universita di Lecce, Lecce,
                      Italy.

Butler D.   Department of Astronomy, Box 2023, Yale University,
                      New Haven, CT 06520, U.S.A.

Carbon D.   Kitt Peak National Observatory, P.O. Box 26732,
                      Tucson, AZ 85726, U.S.A.

Caughlan G.R.*   Department of Physics, Montana State University,
                      Bozeman, MO 59715, U.S.A.

Churchwell E.*   Max Planck Institut für Radio Astronomie, Auf dem
                      Hugel 69, 53 Bonn 1, West Germany.

Clayton R.N.*   Enrico Fermi Institute, University of Chicago,
                      Chicago, IL 60637, U.S.A.

Dearborn D.D.*   Institute of Astronomy, Madingley Road,
                      Cambridge, GB3 OHA, U.K.

Demarque P.*   Department of Astronomy, Box 2023, Yale University,
                      New Haven, CT 06520, U.S.A.

Dickman R.L.   Aerospace Corporation, P.O. Box 92957, Los Angeles,
                      CA 90009, U.S.A.

Encrenaz P.J.*   Radio Astronomie, Observatoire de Meudon,
                        92190 Meudon, France.

Kraft R.P.*   Lick Observatory, University of California,
                        Santa Cruz, CA 95064, U.S.A.

Langer W.D.   G.S.F.C., Greenbelt, MD 20771, U.S.A.

Lequeux J.*   Radio Astronomie, Observatoire de Meudon,
                        92190 Meudon, France.

McCutcheon W.H.   Department of Physics, University of British
                        Columbia, Vancouver, B.C., VGT 1WS, Canada.

Nocar J.L.   Lick Observatory, University of California,
                        Santa Cruz, CA 95064, U.S.A.

Panagia N.   Laboratorio di Astronomia, Via Irnerio 46,
                        40126 Bologna, Italy.

Prialnik D.   Department of Astronomy, Tel Aviv University,
                        Ramat Aviv, Israel.

Rocca-Volmerange B.   Laboratoire René Bernas du C.S.N.S.M., B.P. 1,
                        91406 Orsay, France.

Rolfs C.*   Institut für Kernphysik, Von Esmarchstrasse 10a,
                        44 Munster, West Germany.

Shara M.M.   Department of Astronomy, Tel Aviv University,
                        Ramat Aviv, Israel.

Shaviv G.*   Department of Astronomy, Tel Aviv University,
                        Ramat Aviv, Israel.

Shuter W.L.M.*   Department of Physics, University of British
                        Columbia, Vancouver, B.C., VGT 1WS, Canada.

Snell R.   Astronomy Department, University of Texas at
                        Austin, Austin, TX 78712, U.S.A.

Sparks W.M.   G.S.F.C., Code 671, Greenbelt, MD 20771, U.S.A.

Starrfield S.*   Department of Physics, Arizona State University,
                        Tempe, AZ 85281, U.S.A.

Steigman G.*   Department of Astronomy, Yale University, Box 2023,
                        New Haven, CT 06520, U.S.A.

Tinsley B.M.*   Department of Astronomy, Yale University, Box 2023,
                       New Haven, CT 06520, U.S.A.

Truran J.W.*   Department of Astronomy, The University of Illinois,
                       Urbana, IL 61801, U.S.A.

Tull R.   Astronomy Department, University of Texas at
                      Austin, Austin, TX 78712, U.S.A.

Vanden Bout P.A.*   Astronomy Department, University of Texas at
                      Austin, Austin, TX 78712, U.S.A.

Vigroux L.   Laboratoire René Bernas du C.S.N.S.M., B.P. 1,
                      91406 Orsay, France, and
                      DphEP-ES, CEN de Saclay, 91190 Gif-sur-Yvette.

Vogt S.   Astronomy Department, University of Texas at Austin,
                      Austin, TX 78712, U.S.A.

Walmsley C.M.   Max Planck Institut für Radio Astronomie, Auf dem
                      Hugel 69, 53 Bonn 1, West Germany.

Wannier P.G.*   Radio Astronomy, California Institute of Technologı
                      Pasadena, CA 91125, U.S.A.

Watson W.D.*   Department of Physics, The University of Illinois,
                      Urbana, IL 61801, U.S.A.

Wilson T.L.*   Max Planck Institut für Radio Astronomie, Auf dem
                      Hugel 69, 53 Bonn 1, West Germany.

Winnewisser G.   Max Planck Institut für Radio Astronomie, Auf dem
                      Hugel 69, 53 Bonn 1, West Germany.

PART I

INTRODUCTION

THE IMPORTANCE OF CNO ISOTOPES IN ASTROPHYSICS

Jean AUDOUZE
Radio Astronomie, Observatoire de Meudon, Meudon, France
and Laboratoire René Bernas du Centre de Spectrométrie
Nucléaire et de Spectrométrie de Masse, 91406 Orsay, France.

Carbon, nitrogen and oxygen are the most abundant nuclei after hydrogen and helium (fig. 1) in the observable universe. In fact CNO are elements which are more important than Fe for instance. Historically they have been more difficult to observe in stellar spectra but in spite of these difficulties which have been solved with the development of improved observation techniques and because of their large abundance, as it will be emphasized all throughout this book they play a major role in modern astrophysics. Their observation in the solar system (earth, meteorites, planets, comets, lunar rocks), the stellar spectra, the interstellar medium ... are more and more numerous and precise. More and more information regarding these elements can therefore be gathered and analyzed. This book and the session on which it is based attempt to summarize and to give some hints on all the data which are accumulated at the present on these elements focusing in particular on their isotopic composition. Study of isotopes is indeed most profitable in the sense that they are less affected than elements by chemical fractionation processes although such processes can be considered and be important (see e.g. Watson in this book).

The session organized on the topic of CNO isotopes in astrophysics has tried to give the latest developments on this question in three different directions. (i) The account of the latest abundance determinations in particular those which concern minute but very informative anomalies in the solar system (Clayton), observations of C,N,O abundances in old stars (Demarque, Carbon et al.), and of interstellar molecules. About forty different molecules have been discovered in the interstellar medium,the majority of which being organic : the large chemical activity of carbon explains the richness of the organic chemistry on which all the biology is based. (Wannier, Encrenaz, Wilson et al., Van den Bout et al., Shuter et al., Churchwell et al.). Some of these determinations have been discussed in terms of possible fractionation processes by several contributors such as Shuter and Churchwell but mainly by Watson. (ii) The formation processes i.e. the nucleosynthesis of such elements and isotopes have been recently

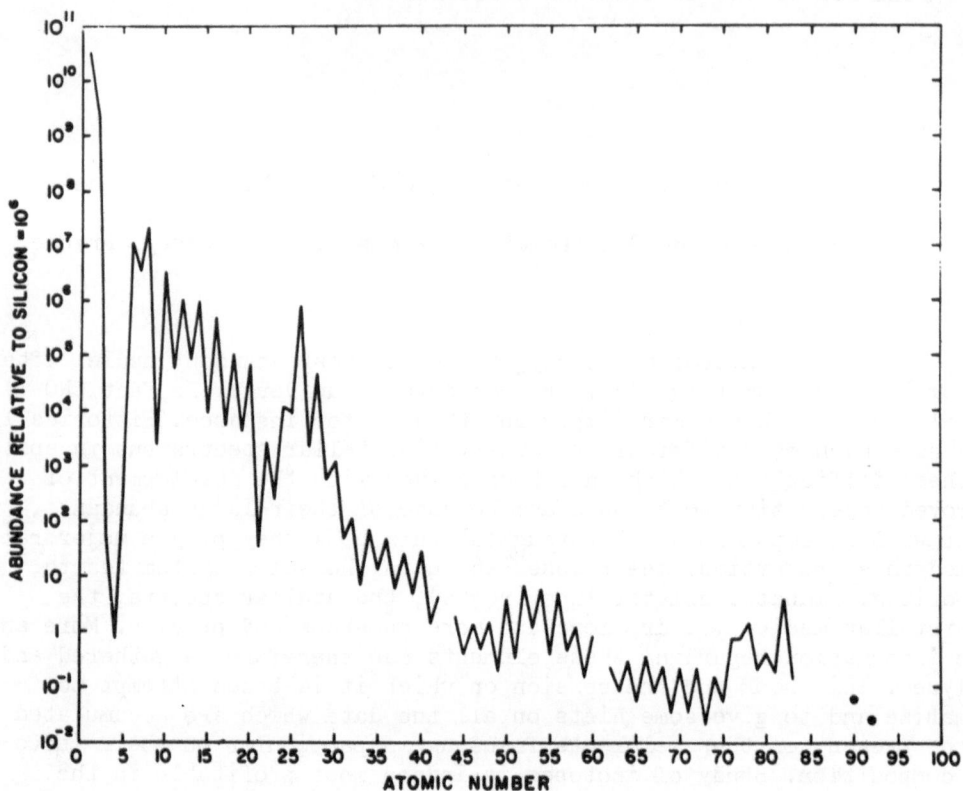

Figure 1 – Curve of the element abundances the atomic number
             (from Cameron 1973, co. the University of Texas Press)

investigated by various authors (see e.g. the review of Truran, but also
the contributions of Caughlan and Starrfield et al). Although these in-
vestigations have been conducted since a rather long time (H. Bethe
has earned the Physics Nobel prize mainly for his contribution to the
understanding of the so called CNO cycle in 1938), new insights has been
obtained in this domain especially due to the outcome of many nuclear
physics experiments relevant to this problem and reported here by Rolfs.
(iii) Finally the evolution of large astrophysical entities such as
galaxies in particular our own Galaxy might be understood by using the
element abundances and isotopic ratios as tracers of these evolution
processes. In galaxies there is a continuous feed-back between the star
formation processes, the ejection of matter during the stellar evolutionary phases and the content of the interstellar gas. Contributions
of Dearborn and Audouze et al deal with this important by-product of
the CNO isotope studies.

     As it can be seen from this program this research includes many
different subfields of astrophysics such as meteoretical studies,
experimental and theoretical nuclear astrophysics, optical astronomy,
radio astronomy ... The purpose of this introductory chapter is to

TABLE 1

Observed isotopic ratios of C, N, and O from Audouze et al., 1975, references are given in this paper – Co. Astronomy and Astrophysics.

Ratio	Interstellar medium				Solar system			
	Ratio	Location	Method	Authors	Ratio	Location	Method	Authors
$\frac{^{12}C^{18}O}{^{13}C^{16}O}$	0.080	14 clouds	CO	Wannier et al., 1975	0.178	Earth		Wedepohl, 1969
	0.10±.01	Sgr B 2	H_2CO	Gardner et al., 1971	0.20±.06	Sun	CO	Hall, 1973
$^{12}C/^{13}C$	36±5	Sgr B 2	HC_3N	Gardner et al., 1975				
	36±8	Ori A	H_2CO 2 mm	Wannier et al., 1975	89±4	Earth		Wedepohl, 1969
	≃36	Ori B	$^{13}C^{18}O$	Lucas, 1975; Wannier et al., 1975				
	≃37	Ori A	CS	Penzias et al., 1972	89±2	Meteorites		Boato, 1954
	22–45	Ori A	HCN	Wannier et al., 1975	≃ 89	Moon		Epstein et al., 1971
	>20	Sgr B 2	H_2CO	Fomalont et al., 1973	90±15	Sun	CO	Hall et al., 1972
	25±5	Sgr A	H_2CO	Fomalont et al., 1973	110±35	Jupiter	CH_4	Fox et al., 1972
	>30	Sgr A	H_2CO	Whiteoak et al., 1974	≃100	Venus	CO	Connes et al., 1968
					≃100	Mars	CO	Kaplan et al., 1969
	12–82	Various clouds	H_2CO	Zuckermann et al., 1974 Evans et al., 1975	70–135	3 Comets	C_2	Danks et al., 1974
	>80	local gas	H_2CO	Whiteoak et al., 1972				
	$42^{+29a)}_{-8}$	ζ Oph	CH^+	Bortolot et al., 1972				
	$72^{+24*)}_{-15}$	ζ Oph	CH^+	Vanden Bout, 1972				
	> 20 to > 77	6 stars	CH^+	Hobbs, 1973				
$^{16}O/^{18}O$	≃390	Sgr A	OH	Wilson et al., 1972	500±25	Earth		Wedepohl, 1969
	>300	Sgr A	OH	Gardner et al., 1970	500±25	Meteorites		Taylor et al., 1965
	>200	Sgr B 2	OH	Gardner et al., 1970	490±25	Moon		Epstein et al., 1971
	385	Ori B	$^{13}C^{18}O$	Wannier et al., 1975	460±150[b]	Sun	CO_2	Hall, 1973
					≃ 500	Venus	CO_2	Connes et al., 1967
$^{17}O/^{18}O$	0.28±.08	Ori A	CO[b]	Encrenaz et al., 1973 Wannier et al., 1975	0.183	Earth		Wedepohl, 1969
	0.25±.13	ϱ Oph	CO[b]	Encrenaz et al., 1973	0.11–0.33[b]	Sun	CO_2	Hall, 1973
	≲0.2	Sgr A, B 2	OH	Zuckermann, 1973				
$\frac{^{12}C^{15}N}{^{13}C^{14}N}$	0.38±.12	Ori A	HCN	Wilson et al., 1972	0.32	Earth		Wedepohl, 1969
	0.28±.06	Ori A	HCN	Wannier et al., 1975				
	0.22	Ori A	HCN	Clark et al., 1974				

[a] Revised values, see text.
[b] Indirect measurement. What is actually measured is $^{12}C^{18}O/^{13}C^{16}O$ and $^{12}C^{17}O/^{12}C^{18}O$ (or $^{12}C^{15}O/^{13}C^{16}O$).

give some overview of the topic and guideline among these different
subfields and in particular take an opportunity to mention some
important contributions which have not been explicitly reported in this
short session.

I - THE ABUNDANCES AND ISOTOPIC RATIOS

Observations have been made in the solar system, in stars and also
in the interstellar medium.

Figure 1 shows the universal abundances (standard abundance distribution based on a comparison between solar and meteoretical abundances made by Cameron, 1973). This distribution shows that C, N and O
are the elements which are the most abundant after H and He. Besides
their large abundance the interest of these elements is that their
isotopic ratios have been observed in various astrophysical locations.
Table 1 gathers some of the observed isotopic ratios of C, N and O in
the solar system and in the interstellar medium (as deduced from
molecular determinations) and Fig. 2 shows the $^{12}C/^{13}C$ ratio observed
at the surface of some giant stars. More information is indeed given in
this book especially regarding the interstellar ratios derived from
molecular determinations. From the available data, several important
conclusions can be extracted :

1 - If we except small but definite anomalies in the oxygen or nitrogen
isotopic ratios as determined for instance by Clayton (this book) in
meteorites or in the lunar samples (the magnitude of these anomalies is
about a few percent), the isotopic composition in CNO seem to be constant
in the whole solar system (earth, planets, solar surface, meteorites).
Since comets are found to have solar system $^{12}C/^{13}C$ ratios, unfortunately
within rather large uncertainties, their solar origin seems to be very
likely.

2 - The anomalies in oxygen isotopic ratios have been determined in
mineralogical fractions of some carbonaceous chondrites (e.g. Allende)
by Clayton and his associates. Other isotopic anomalies have
been determined in carbonaceous chondrites for Ne, Mg, Xe... These
anomalies which are shortly discussed in this book are tentatively
explained in terms of admixture of extra solar system material which
has experienced a different nucleosynthetic history.

3 - C, N, O abundances have been observed in different types of stars :
(i) there exists a class of stars : the carbon stars which have been
thoroughly reviewed by Wallerstein (1973) in which C/H can be larger
than 1 (like in RCrB, RYSgr or HD 182040). In RYSgr ^{14}N and ^{16}O are not
as abundant as ^{12}C but have an abundance comparable to H. (ii) In population II stars carbon and nitrogen abundances seem to vary largely in
stars belonging to gobular clusters (see in this respect the contributions of Demarque and Carbon et al). In particular the nitrogen abundance seems to vary strongly from one object to the other this variation
is stronger for old population II stars than for the stars belonging
to the disk population. (iii) The ^{13}C isotope is enriched relative to
^{12}C in many giant stars the enrichment being correlated with the stellar
luminosity (fig. 2). Giants are supposed to eject material enriched

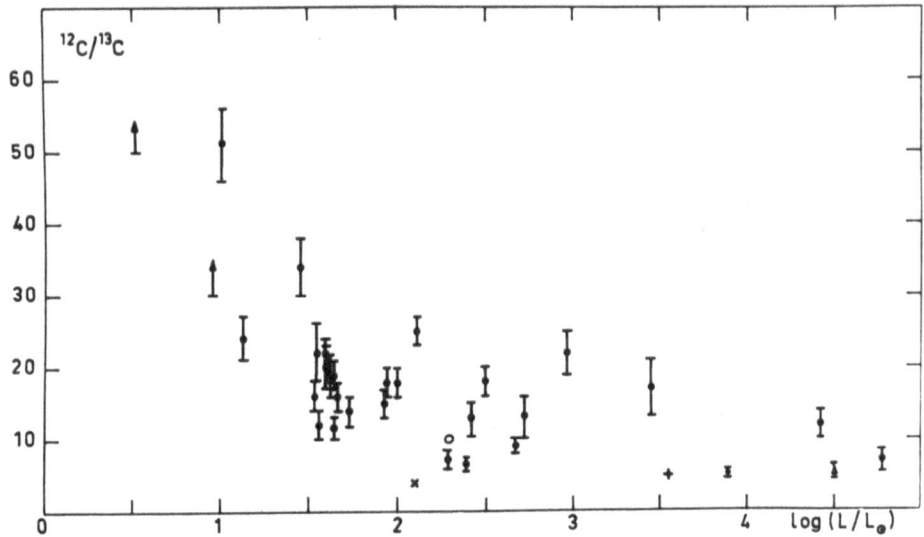

Figure 2 – $^{12}C/^{13}C$ ratios observed in giant stars against the stellar
luminosity.
From Vigroux et al, 1976, Co. Astronomy and Astrophysics.

into ^{13}C (see e.g. the review of Dearborn). (iv) In one carbon star and
two red giants ^{17}O appears to be enriched relative to ^{16}O (as recalled
by Encrenaz). (v) Planetary nebulae represent the latest stage of the
evolution of low mass stars. These objects appear to be enriched into
C and N (see e.g. Panagia et al). (vi) Finally, novae which are explosive
objects able to release energy of $\sim 10^{42-44}$ ergs are enriched into N,
and O (and presumably C) (the latest observations of these objects are
those of Andrillat and Collin-Souffrin, 1976). They appear to be enriched into ^{13}C and ^{15}N (see the work of Sneden and Lambert on DQ Hercul
observations, 1974).

These observations are taken into account in the models of nova
outbursts by Starrfield et al, and in some extent (for the dwarf novae)
in the models of Prialnik et al.

4 – The CNO composition has been thoroughly studied in the interstellar
medium. (i) In the solar neighborhood, optical observations in the
ultraviolet wavelength range have been made by the Princeton group (see
the review of Spitzer and Jenkins, 1975) from which C, N, O may be
underabundant with respect to their solar system value by a factor 3-5.
It must be remembered that the depletion of all the complex elements
(heavier than helium) in the interstellar gas is presumably due to the
fact that these elements form the interstellar dust (see e.g. Field,
1974). Furthermore studies of the X-ray absorption due to interstellar
medium and which integrate on the solid + gaseous phase show that C, N

and O have a normal abundance in the interstellar medium (Ryter et al, 1975). (ii) N, and O abundances exhibit a spatial gradient in nearby spiral galaxies M 101 and M 33 for instance, and almost certainly in our own Galaxy : the oxygen abundance increases by factors of $\sim$ 10 from the outer to the inner regions (regions closer to the center) while the nitrogen abundance increases by factors as large 30-40 : the N/O ratio increases in the central regions of the spiral galaxies. (iii) the interstellar gas is at least apparently enriched into ^{13}C (relative to ^{12}C) by factors around 2 (with respect to its solar system value) ; the enrichment seems to be slightly larger in the central regions than in the solar neighborhood. A large fraction of this book is devoted to the report on the last measurements dealing with this possible enrichment and to the analysis of their implications in terms of chemical evolution effects as present (e.g. Audouze et al) but keeping in mind possible fractionation effects such as those discussed here by Watson. In this respect nitrogen and oxygen isotopic ratios have a value different from their solar system value (enrichment of ^{17}O with respect to ^{18}O in the central regions of our galaxies as recalled by Encrenaz and depletion of the $^{12}C^{15}N/^{13}C^{14}N$ ratio as measured by Linke et al, 1977). Such variations are going to be interpreted in terms of evolution (see Audouze et al).

5 - C, N, O nuclei have been detected in the galactic cosmic rays. The main features of the cosmic ray composition is that C and O have roughly the same abundances (in fact C seems to be slightly more abundant than O). Furthermore the galactic cosmic ray composition is enriched into ^{13}C and ^{15}N relative (respectively) to ^{12}C and ^{14}N. The rough equality of C and O abundance (contrary to C/O $\sim$ 1/3-4 in the solar system) might be due either to an enrichment of the galactic cosmic ray sources in ^{12}C (relative to ^{16}O) which would mean that these sources are supernovae with an average mass of $\sim$ 10 M_O or to the difference between the first ionization potential (^{12}C being more easily ionized than ^{16}O) which would mean that C is slightly more easily accelerated than O during the release of galactic cosmic rays by any active star (this last hypothesis is favoured by Cassé, 1976). The enrichment of galactic cosmic rays into odd nuclei such as ^{13}C and ^{15}N has the same cause as the enrichment into light elements LiBeB/CNO)$_{GCR}$ $\sim$ 0.23 while LiBeB/CNO)$_{SAD}$ $\sim$ 10^{-5} or iron peak nuclei such as V, Cr or Mn : during the propagation of galactic cosmic rays throughout the interstellar matter some ^{13}C and ^{15}N are produced by spallation reactions such as $^{14}N(p, pn)$ ^{13}N ( ) ^{13}C and $^{16}O(p,pn)$ ^{15}O (β^+) ^{15}N induced during this interaction. The difference between the $^{13}C/^{12}C$ and $^{15}N/^{14}N$ ratio in the cosmic rays and in the standard distribution allows to measure the cosmic ray path through the interstellar medium which is found to be comparable (X $\sim$ 6 g cm^{-2}) to that deduced from the light element (LiBeB) measurements.

II - NUCLEOSYNTHESIS AND EVOLUTION OF GALAXIES

This introduction will be shorter on these two points since they are extensively covered below in papers such as those of Caughlan, Dearborn, Starrfield et al, Truran, Audouze et al, and Tinsley. The important features on which the attention must be drawn are the following:

1 - ^{12}C and ^{16}O are amongst the stablest nuclei ^{16}O is a double magic number nucleus (Z = N = 8) ^{14}N is the last stable odd-odd nucleus (after D, ^{6}Li and ^{10}B). In evolved supergiants ^{12}C and ^{16}O can induce C + C and O + O fusion reactions when the temperature is larger than $6 \cdot 10^{8}K$ for carbon burning and $10^{9}K$ for oxygen burning. These processes explain the formation of elements between $20 \leqslant A \leqslant 30$. When these processes occur explosively at higher temperatures (T $\simeq 2 \cdot 10^{9}K$ for carbon burning and T $\simeq 3 \cdot 10^{9}K$ for oxygen burning) they account rather satisfactorely for the observed isotopic ratios between neon up to the iron peak (see e.g. the review of Arnett, 1973). Finally the explosive fusion of carbon might occur in the degenerate core of $4 < M/M_{0} < 8$ stars and induce their explosion. The drawback of this mechanism which has been examined with the purpose of explaining the supernova explosions is that the explosion would occur without leaving any remnant contrary to what is observed in an actual supernova outburst. Furthermore if this mechanism occurs for all stars in this mass range the universe would have a Fe abundance much higher than what is actually observed (see e.g. in this respect Arnett, 1969).

2 - Although there are only seven different stable isotopes ^{12}C, ^{13}C, ^{14}N, ^{15}N, ^{16}O, ^{17}O and ^{18}O more than one nucleosynthetic process must be invoked to explain their formation in the present stage of the available theories on nucleosynthesis, about four different processes must be invoked : 1) the cold CNO cycle occuring in the hydrogen burning zone of main sequence stars or red giants explains the ^{14}N and also when the CNO cycle is incomplete (n(H) $\lesssim$ n(C)) the ^{13}C formation. 2) the helium burning in red giant stars explains the ^{12}C and ^{16}O formation and possibly that of ^{18}O through the reaction $^{14}N(\alpha, \gamma)^{18}F$. The explosive nucleosynthesis in novae, supernovae or in the still hypothetical supermassive stars would explain the formation of ^{13}C, ^{15}N and ^{17}O if the explosive processes occur in hydrogen rich zone and that of ^{15}N and ^{18}O if the explosive processes occur in helium rich zone. Although a lot of thought and work has been devoted to the nucleosynthesis of these nuclei, there is no hope at present to find out a single and unitary process to explain the seven CNO isotopes in the same event.

3 - In nuclear astrophysics one often distinguishes between primary and secondary elements : the primary elements are those which can be theoretically formed in a metal free star directly from a hydrogen-helium gas suffering a succession of gravitational and nuclear burning phases without intense mixing processing. Of the seven CNO isotopes ^{12}C and ^{16}O are primary elements which come directly from the fusion of helium. Note : ^{14}N can also be considered in some circumstances as a primary element if there is sufficient mixing between the hydrogen burning zone and the helium burning zone of the H-He star. On the contrary secondary elements need a metal (Z > 2) seed to be formed. That is the case of ^{13}C, ^{14}N, ^{17}O and ^{18}O which are made from ^{12}C and ^{16}O in the various processes recalled above.

This distinction is not purely semantic. In the time-evolution of the galaxies secondary elements are expected to have stronger variations in their abundances than primary elements (see e.g. Audouze and Tinsley, 1976). In fact that is what one observes in our Galaxy and in nearby galaxies : as soon as the galactic material is more strongly processed into stars,one expects an increase in the secondary element abundances relative to those of the primary elements. The major parameters in models of chemical evolution of galaxies are the rate of star formation from the interstellar material and the rate with which they release the processed material either by mass loss or by their quiet or explosive death. These processes depend strongly on their initial mass. C, N, O isotopes which have very different nucleo-synthetic histories and which are therefore produced in different stellar regions or during different phases of the stellar evolution are good indicators of the way by which galaxies evolve. That is the reason why so much attention has been paid to CNO isotopes in the frame of such evolution models.

4 - In this respect, elements such as ^{13}C and ^{14}N have an interesting feature in contrast with elements such as ^{12}C and ^{16}O : they can be manufactured in low mass stars in the red giant or in the planetary nebula phase. Therefore they can be used as tracers of low mass star nucleosynthesis compared to the overall metal abundance which should be the signature of high mass star processing. In particular since mole-cular clouds have been observed in the central regions of our Galaxy some information on the evolution of such regions can be extracted from the isotopic measurements of these clouds.

In the following chapters in particular in that by Watson, several words of caution will be made regarding the use of this data in terms of possible chemical fractionation and on the ground that the solar system might not be as typical of the overall isotopic landscape as one would like.

All throughout this book it is stressed by the different contri-butors and especially well summarized by Tinsley how much patchy and sometimes confusing our ideas can be about the astrophysic studies of these isotopes. The goal of this book in which many different and not yet settled ideas, hypotheses and theories are presented and confronted will be achieved if it prompts scientists interested in different fields to use their expertise in this multidisciplinary question : various burning problems such as the physics of the interstellar medium, the origin of the solar system, the nucleosynthesis of CNO elements, the stellar evolution and the evolution of the solar neighborhood and the galactic center might be enlightened soon if effort is made in this area of nuclear astrophysics.

At the end of the edition of this book, I learnt the dreadful accidental decease of Dr. Mireille BERTOJO, who was 24, and has authored the pioneering review paper on the isotopes anomalies in C, N and O observed in the interstellar medium (M. Bertojo, M.F. Chiu and C.H. Townes, 1974, Science, 184, 619). Let me dedicate my introductory remarks to the memory of our bright young colleague.

References

Andrillat, Y. and Collin-Souffrin, S., 1976, in press (see also
    Collin-Souffrin, S., 1977, in Novae, M. Friedjung ed., Reidel
    Publ. Co. in press).
Arnett, W.D., 1969, Astrophys. Space Sci.
Arnett, W.D., 1973, Ann. Rev. Astron. Astroph., 11, 73.
Audouze, J. and Tinsley, B.M., 1976, Ann. Rev. Astron. Astroph., 14, 43.
Cameron, A.G.W., 1973, in Explosive Nucleosynthesis (D.N. Schramm and
    W.D. Arnett ed., The University of Texas Press) p. 3.
Cassé, M., 1976, unpublished PhD Thesis (Université de Paris VII).
Field, G.B., 1974, Ap. J., 187, 453.
Linke, R.A., Goldsmith, P.F., Wannier, P.G., Wilson, R.F. and Penzias,
    A.A., 1977, Ap. J., in press.
Ryter, C., Cesarsky, C.J. and Audouze, J., 1975, Ap. J., 198, 103.
Sneden, C. and Lambert, D.L., 1975, M.N.R.A.S., 170, 533.
Spitzer Jr., L. and Jenkins, E.B., 1975, Ann. Rev. Astron. Astroph.,
    13, 133.
Wallerstein, G., 1973, Ann. Rev. Astron. Astroph., 11, 115.

References

Amaïllov, ... and Villa-Brence, ... 1972 ...
...

PART II

CNO ISOTOPES IN THE SOLAR SYSTEM

1) OXYGEN ANOMALIES IN METEORITES
2) THE $^{15}N/^{14}N$ RATIO IN THE SOLAR WIND[+]

Robert N. CLAYTON
Enrico Fermi Institute, The University of Chicago

I - OXYGEN ANOMALIES IN METEORITES

The meteorites have long served as the most diverse collection of samples of the solar system available for detailed analysis in the laboratory. Their isotopic composition has been thoroughly studied recently. Indeed as we will show here, they may provide some insight on the way that material which has formed the solar system has been homogenized.

Isotopes anomalies have been serached in meteorites for various elements (Table 1). The enrichment in ^{13}C is almost certainly of chemical origin. But in this list the chemical elements which have three

TABLE 1

Summary of enrichment effects observed in some meteorites

Isotope	Enrichment (%)	Meteorites where it has been observed
^{13}C	7	Carbonates in C1 and C2 meteorites
^{15}N	17	Renazzo (C2)
^{16}O	5	High temperature phase in C2-C3 meteorites
^{22}Ne	< 1	Low temperature phase in C1-C2
^{26}Mg	1.3	Aluminium rich phase in Allende (C2)

[+]This excerpt of the talk of Dr. Clayton has been written up by the editor.

and more stable isotopes can be used to discriminate between chemical fractionation and nuclear processes effects. In this respect an extensive study of the oxygen anomalies has been found very useful to determine the degree of inhomogeneity of the solar system material. Figure 1 which is extracted from a paper of Clayton and Mayeda (1976) is a three-isotope plot of the $^{17}O/^{16}O$ variation with respect to the $^{18}O/^{16}O$ variation.

In this figure the parameters δ are (for instance for δ^{18})

$$\delta^{18} = \frac{(^{18}O/^{16}O) \text{ sample}}{(^{18}O/^{16}O) \text{ SMOW}} - 1 \quad x \; 1000$$

Two lines can be distinguished on this figure :

The first line has a slope 1/2 (i.e. $\delta^{17} = 1/2 \; \delta^{18}$) and gathers the isotope anomalies observed in terrestrial, lunar and meteoritical samples ; this line is well explained in terms of chemical fractionation

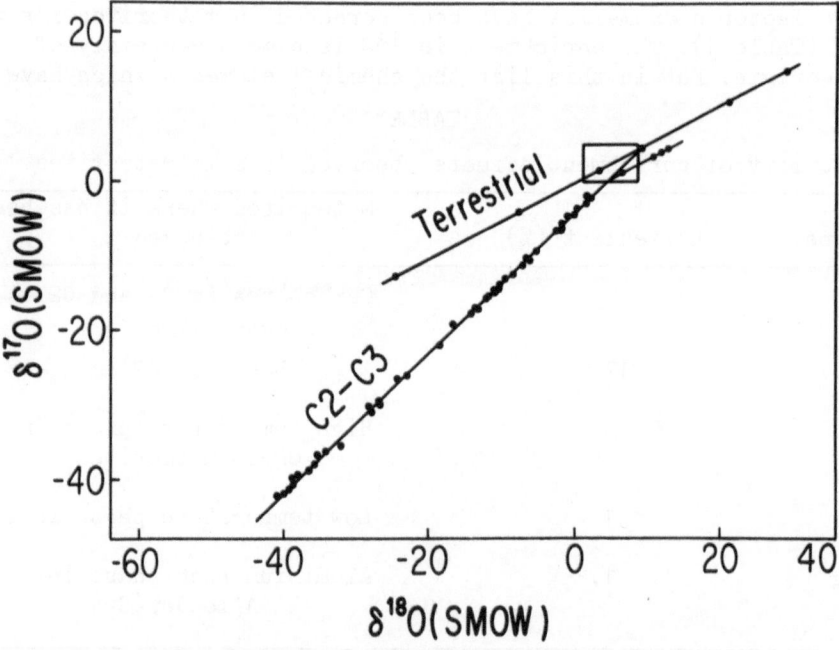

Figure 1 – Three isotope plot illustrating large scale variations in oxygen isotopic abundances. 1) a mass fractionation line defined by several terrestrial materials. 2) a mixing line defined by anhydrous phases of C2 and C3 meteorite, resulting from admixture of an ^{16}O-rich component.

SMOW = standard mean ocean water (from Clayton and Mayeda, 1976)

the effects on the mass 17 should be half of those affecting the mass
18. The second line shows ^{17}O anomalies much lower than those of the
first line. The second line gathers determinations which have been made
from anhydrous phases of C2 and C3 carbonaceous chondrites (in particular
the large white inclusions of the C2 Allende meteorite). The slope of
this line is 1 (i.e. $\delta^{17} \sim \delta^{18}$). This has been interpreted by Clayton,
Grossman and Mayeda (1973) in terms of different nuclear processing :
the admixture of 5 % of a pure ^{16}O phase in these meteoritical components
would explain this line with a slope of 1.

The interest of such a discovery is three fold :

1 - The solar system material is indeed not fully homogeneized.
Similar isotope anomalies effects have been found in particular for
elements listed in Table 1.

2 - The oxygen anomaly (consisting of an admixture of a pure ^{16}O phase)
has been interpreted by the presence in the inclusions of C2 and C3
meteorites of interstellar grains which would have a different nucleosynthetic history : If some interstellar grains have been formed in the
vicinity of an exploding object such as a supernova (or if their formation have been triggered by such an explosion) their isotopic composition should have been affected by the nucleosynthesis which has
accompanied such explosions. In particular one can accept the idea that
grains formed close to a supernova remnant are enriched into ^{16}O which
comes from the helium burning reactions. Several contributions such as
the recent work of Cameron and Truran (1976) have attempted to give a
single explanation to all the observed anomalies in terms of a correlation
with nucleosynthesis occuring near the regions where such grains have
been formed. (see also D.D. Clayton, 1975).

3 - Clayton and Mayeda (1976) have used this inhomogeneous distribution
of the oxygen isotopes to identify solar system bodies which were
formed from a common region of the solar nebula and to distinguishthem
from bodies formed in other regions. In particular they show that the
moon belongs to the same group as the earth and the differentiated
meteorites such as the achondrites, the stony-iron and the iron meteorites while it is unrelated to the carbonaceous achondrites such as the
ordinary chondrites. Clayton and Mayeda (1976) interpret this by assuming
that the moon has been more likely formed by accretion of bodies which
have previously undergone an igneous differentiation preferably to a
direct formation from nebular condensates.

II - THE ^{15}N/^{14}N RATIO IN THE SOLAR WIND

Nitrogen is not as favourable as oxygen to separate anomalies
between those due to chemical effects and those due to nuclear processing
since it has only two stable isotopes ^{14}N and ^{15}N, preventing any
analysis such as that presented in § 1. However the ^{15}N/^{14}N ratio has
been extensively studied in lunar samples and can be used to determine
the ^{15}N/^{14}N ratio of the solar wind : as noted by several authors (see
e.g. Kerridge, 1975) the presence within lunar soils of C, N, H and
noble gases which are effectively absent from the lunar rocks should

be due to the implantation of solar wind in the surface of lunar grains. What has been done by Becker and Clayton 1975, and also by Kerridge et al. 1975, was to analyse the $^{15}N/^{14}N$ ratio in the gas implanted in lunar samples while there is little nitrogen implanted in the rocks and the breccias (less than 5 ppm) ; soils and lunar fines yield up to 40-120 ppm nitrogen. There is indeed an anticorrelation between the nitrogen content and the size of the grains.

From the experiments reported for instance by our group (Becker and Clayton, 1975) and based on step-wise heating techniques, large ^{15}N anomalies have been found in the solar wind implanted in lunar fines : the $^{15}N/^{14}N$ ratio has been found $\sim$ 12 % larger in such samples than in terrestrial atmosphere. Our experiments allow to separate the $^{15}N/^{14}N$ ratio coming from the spallation of lunar rocks (this fraction is released at very high temperature by step-wise heating techniques because the spallogenic nitrogen is well fixed inside the mineralogic crystals).

May be the most important finding coming from these measurements is the fact that the $^{15}N/^{14}N$ ratio coming from the solar wind shows a secular increase in the $^{15}N/^{14}N$ ratio. This finding has also been made by Kerridge (1975) who has noted that the $^{15}N/^{14}N$ ratio has increased by about 15 % within the last 3 to 8 10^8 years. From our work we find that δN^{15} = - 9.5 % (relative to the terrestrial value) and for the present solar wind δN^{15} = + 12 % which corresponds to a secular increase of 20 % for about 3 10^8 years. This conclusion comes mainly from the fact that the large (δN^{15}) anomalies are related with low Ne^{21} age (Fig. 2) (the decrease appear to be roughly linear with respect to this age in Kerridge, 1975). This result means that either the whole photosphere or the solar events which generate preferentially ^{15}N have been enriched into ^{15}N relative to ^{14}N. The origin of such an increase is still a mystery. In any case it cannot be due to mass-dependant fractionation processes otherwise one should see larger effects on carbon and the $^4He/^3He$ ratio.

Such an increase seems presently difficult to be explained in terms of spallation reactions (this should go together with a strong effect on light elements Li Be B and especially on boron which should be more easily formed than ^{15}N. Similarly nuclear reactions occuring inside the sun have the net result of decreasing ^{15}N with respect to ^{14}N as shown by Boschler and Geiss (1973). Some tentative explanations could be (i) a change of the surface composition of the sun but one does not see any mechanism able to do it easily. Furthermore this mechanism should affect the isotope composition of other elements : this has not yet been seen. (ii) this solar wind nitrogen might be formed in special solar flares events:one may recall at this point the very strange $^3He/^4He$ ratios sometimes larger than 1 found in many low energy flares as seen by for instance the Caltech group (Hurford et al., 1975) and the Chicago group (Anglin, 1974). May be solar flare scenarios as peculiar as those presented by Colgate et al., 1977, to account for such strange isotopic composition of H and He in these flares might explain the

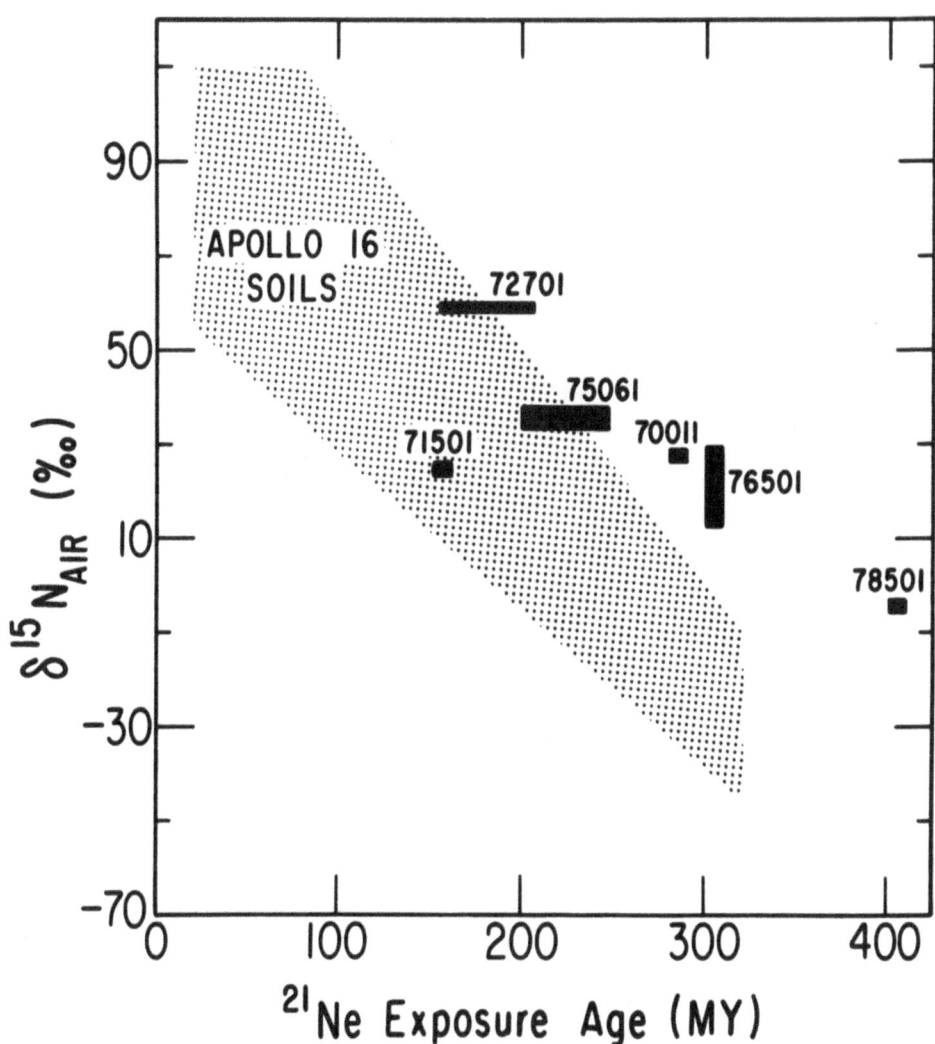

Figure 2 - Nitrogen isotope ratios and ^{21}Ne exposure age for
Apollo 16 and 17 bulk soil samples.

From Becker and Clayton.

increase of the $^{15}N/^{14}N$ ratio. Finally as noted by Becker and Clayton, 1975, this increase in ^{15}N might be the result in the change of the $^{15}N/^{14}N$ ratio coming from the decreasing contribution of an isotopically light, indigeneous lunar-nitrogen component outgassed from the interior of the moon and reimplanted by solar wind ionization and acceleration into the lunar surface.

The $^{15}N/^{14}N$ ratio has been also investigated in meteoritical samples : this ratio is similar to that found in the terrestrial samples (within 4 %). However, in the C2 carbonaceous chondrite Renazzo an increase of $^{15}N/^{14}N$ of 17.2 % has been found. At present one does not know whether such a big enerichment is due to chemical effects or to nucleosynthetic processes.

These two examples,the oxygen anomalies in meteorites and the study of the $^{15}N/^{14}N$ ratio increasing secularly in the solar wind, show how much information can be grasped from these studies which shed invaluable light not only on the evolution of the solar system but possibly also on the nucleosynthetic history of the solar system and some parts of the interstellar matter.

REFERENCES

Anglin, J.D., 1974, Ap. J. Letters, 186, L 41.
Becker, R.H. and Clayton R.N., 1975, preprint.
Boschler, P., and Geiss, J., 1973, Sol. Phys., 32, 3.
Cameron, A.G.W. and Truran, J.W., 1976, preprint.
Clayton, D.D., 1975, Nature, 257, 36.
Clayton, R.N., Grossman, L. and Mayeda, L.K., 1973, Science, 182, 485.
Clayton, R.N. and Mayeda, T.K., 1976, preprint.
Colgate, S.A., Audouze, J. and Fowler, W.A., 1977, Ap. J., in press.
Hurford, G.J., Mewaldt, R.A., Stone, E.C. and Vogt, R.E., 1975,
    Ap. J., 201, L95.
Kerridge, J.F., 1975, Science, 188, 162.
Kerridge, J.F., Kaplan, I.R. and Petrowski, C., 1975, Geochim.
    Cosmochim. Acta, 39, 137.

REFERENCES

PART III

CNO ISOTOPES IN STARS

POSSIBLE LARGE CARBON AND NITROGEN ABUNDANCE VARIATIONS ON THE
HORIZONTAL BRANCH OF M92

L.H. Auer   and   Pierre Demarque
High Altitude Observatory   Yale University Observatory

ABSTRACT

Model stellar atmospheres have been constructed to investigate the
effects of the continuous opacity of CI, NI and OI on the Balmer jump
of late-B horizontal branch stars. The large opacity of CI and NI in
the ultraviolet raises the Balmer continuum through back warming without changing the Paschen continuum, while OI has little effect. The
models provide an explanation for the variations in the Balmer jump
which have been observed from star to star by Oke near 12,500°K on
the horizontal branch of the globular cluster M92. We conclude that
some stars in M92 appear to have an atmospheric C and/or N content
between one hundred and one thousand times the cluster average. The
same effect of the opacities of CI and NI may also explain Newell's
gap NI in the [(U-B)-(B-V)]-diagram for blue stars in the galactic halo.

I.   INTRODUCTION

We present results of stellar atmosphere calculations to investigate the effect of abundance variations of He, C, N and O on the size
of the Balmer jump in late-B stars on the horizontal branch of the
halo population. The starting point of this project was an attempt to
understand the rather puzzling observations of variations in the
magnitude of the Balmer jump among some blue stars of the same color   .
in M92 made by Oke (1975). A further motivation came from the work
of Newell (1973) on the distribution of blue halo stars in the field
in the [(U-B)-(B-V)]-diagram.

1.   The Balmer jump in M92.

Observations made by Oke (1975) with the multichannel photoelectric
scanner mounted on the 200-inch Hale telescope have revealed the
existence of three horizontal branch stars in M92 which exhibit Balmer
jumps D_B smaller by as much as 20% in comparison with other stars of
the same effective temperature T_{eff} (a total of six stars have been

Jean Audouze (ed.), CNO Isotopes in Astrophysics, 25-32. All Rights Reserved.
Copyright © 1977 by D. Reidel Publishing Company, Dordrecht-Holland.

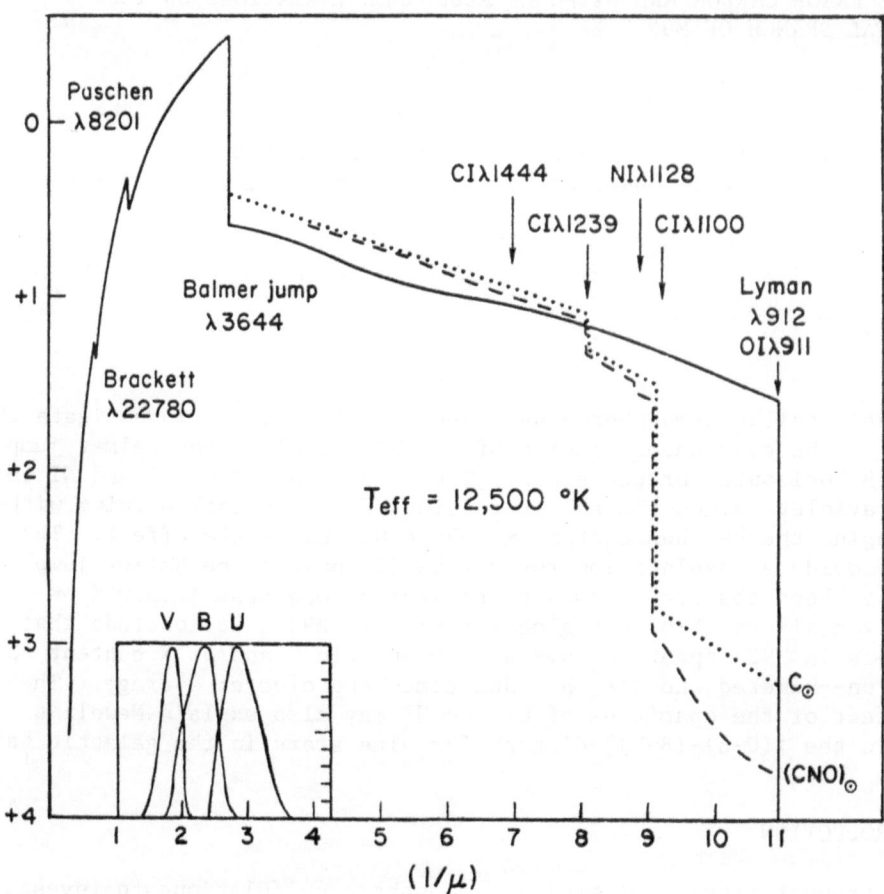

Figure 1. The continuum for three models of the same temperature (T_{eff}=12,500°K) expressed in magnitude vs.
wave number in microns^{-1}. The solid line represents an atmosphere of pure hydrogen and helium.
The dotted line includes the additional effect of
the solar carbon abundance; the dashed line that
of solar (CNO)-abundance. The U,B,V filters are
from Johnson and Morgan (1953).

observed). The Paschen slopes of these stars, however, are very
similar and imply T_{eff}~12,500°K. on the basis of the Mihalas (1964)
model stellar atmospheres.

    2. Newell's gap 1 (N1).

    Quite independently, Newell (1973) has emphasized the presence of
two gaps (N1 and N2) in the distribution of blue halo stars in the
[(U-B)-(B-V)]-diagram. The cooler of the two gaps, N1, coincides

closely in temperature with Oke's anomalous Balmer jumps, and we shall
concern ourselves here only with N1. Newell (1973) found that N1 is
particularly well defined in the field. The existence of N1 has been
confirmed by Greenstein and Sargent (1974). More recently, Newell and
Graham (1976) have also found evidence for N1 in the $[C_1-(b-y)]$-diagram
of the Strömgren photometric system for a sample of blue halo stars.

We believe that the cause of Newell's N1 gap is related to the
same physical process which is responsible for the anomalous Balmer
jumps discovered by Oke in M92. The coincidence in temperature between
the two phenomena is remarkable. Furthermore, in Oke's observations,
the Balmer continuum is raised without a corresponding change in the
Paschen continuum. The implied change in the $[(U-B)-(B-V)]$-diagram
is in exact agreement with the existence of the N1 gap. This is
illustrated in Figure 1 where the spectral response of the U,B and V
filters is shown.

II.   THE MODEL ATMOSPHERES

Model stellar atmospheres were constructed for the range of
effective temperatures ($0.35<\theta_{eff}<0.45$, where $\theta=5040/T$) with the
help of a computer program written by Auer and Heasley (1973). The
program was modified to include the effects of C, N and O opacities
in addition to hydrogen and helium included in the original code. All
models were constructed with the surface gravity log g = 3.7 appropriate
to blue horizontal branch stars. This is not a critical choice since
D_B depends primarily on temperature in the temperature range and little
on surface gravity (Mihalas 1964). The reader is referred to the
paper by Auer and Demarque (1977) for a more complete discription of
the models. In summary, the influence of abundance changes on D_B are
the following:

1.   The helium abundance

The Balmer jump is essentially invariant against changes in
y(the helium abundance relative to hydrogen by number) at these
temperatures. The helium is neutral and therefore changes in its
abundance mimic changes in the surface gravity. We have seen that
D_B is extremely insensitive to surface gravity above 10000°K and it
is thus likewise insensitive to y. Only for y>100 is there any notice-
able effect and values this high appear inconsistent with interior
models at this temperature (Gross 1973; Sweigart, Mengel and Demarque
1974). Our results are in agreement with an earlier investigation by
Böhm-Vitense (1967).

2.   The oxygen abundance.

Oxygen like helium cannot influence the magnitude of the Balmer
jump. The only important absorption by O occurs from the ground state.
This edge is almost exactly coincident with the Lyman edge of hydrogen

and is thus totally masked due to the much larger abundance of hydrogen
in the atmosphere.

3.   The carbon and nitrogen abundances.

The neutral atoms of C and N have strong absorption edges above
the Lyman limit and are extremely effective in blocking flux in the
ultraviolet portion of the Balmer continuum. The effect on D_B is indirect and is illustrated in Figure 1. In order to carry the same
total flux, the atmosphere adjusts itself to carry the blocked flux in
a different region of the spectrum. Since the Lyman continuum is very
opaque, it must be shifted to higher wavelengths in the Balmer continuum
resulting in a decrease in D_B. The Paschen continuum is formed so
deep in the atmosphere that it is unaffected.

Qualitatively and quantitatively, the effect of increasing the
abundances of C and N is very similar. No change in D_B is found for
abundances as low as 10^{-2} solar. As one progressively increases the
abundance to ten times the solar value, the effect then saturates and
actually reverses slightly due to changes which occur in the Paschen
continuum.

The maximum deviation ΔD_B in D_B from the pure H-He atmosphere
occurs near $\theta_{eff}=0.40$ for a mixture of C,N and O and near $\theta_{eff}=0.37$ for
N alone. At maximum, $\Delta D_B \approx 0.12$, requiring CNO-abundances from solar to
ten times the solar value (see Auer and Demarque 1977).

III.   POSSIBLE INTERPRETATION OF THE OBSERVATIONS

1.   The Balmer jump in M92.

Our calculations are consistent with the interpretation that the
variations in D_B observed by Oke (1975) are caused by variations in
C and/or N atmospheric abundances from star to star. Since M92 is one
of the most metal poor globular clusters in the galactic halo, i.e.
[Fe/H]≈-2(Butler 1975), and since an abundance of C and/or N somewhere
between solar and ten times solar seems required to affect D_B,by the
amount needed, our interpretation requires an abundance enhancement by
a factor of one hundred to one thousand over the generally accepted
mean abundance of the cluster.

We must emphasize at the same time that the method provides no
information on abundances of elements other than C or N which do not
affect the size of the Balmer jump. In particular, helium, oxygen, or
the s-process elements may or may not be enhanced in Oke's "small-D_B"
stars.

Neither is it possible with the information at hand to different-
iate between enrichment in C or in N. There is however a hint that N
is the overabundant element. It seems reasonable to conjecture that

the "small-D_B" stars are related to the "weak-CH" (or weak G-band) stars observed on the asymptotic branch of M92 (Zinn 1973). In fact, Butler, Carbon and Kraft (1975) have recently obtained spectral scans of several weak G-band stars at the Lick Observatory which when analyzed exhibited strong enhancement in nitrogen.

2. The presence of N1 in the [(U-B)-(B-V)]-diagram.

We have noted earlier the coincidence between the temperature of N1 (at log $T_{eff} \approx 4.11$) and the temperature of the anomalous Balmer jumps. The nature of N1 can be illustrated by considering a stellar distribution $N(\theta_{eff})$ along the horizontal branch such that:

$$\frac{dN}{dD_B} = \frac{dN}{d\theta}_{eff} \cdot \frac{d\theta_{eff}}{dD_B}$$

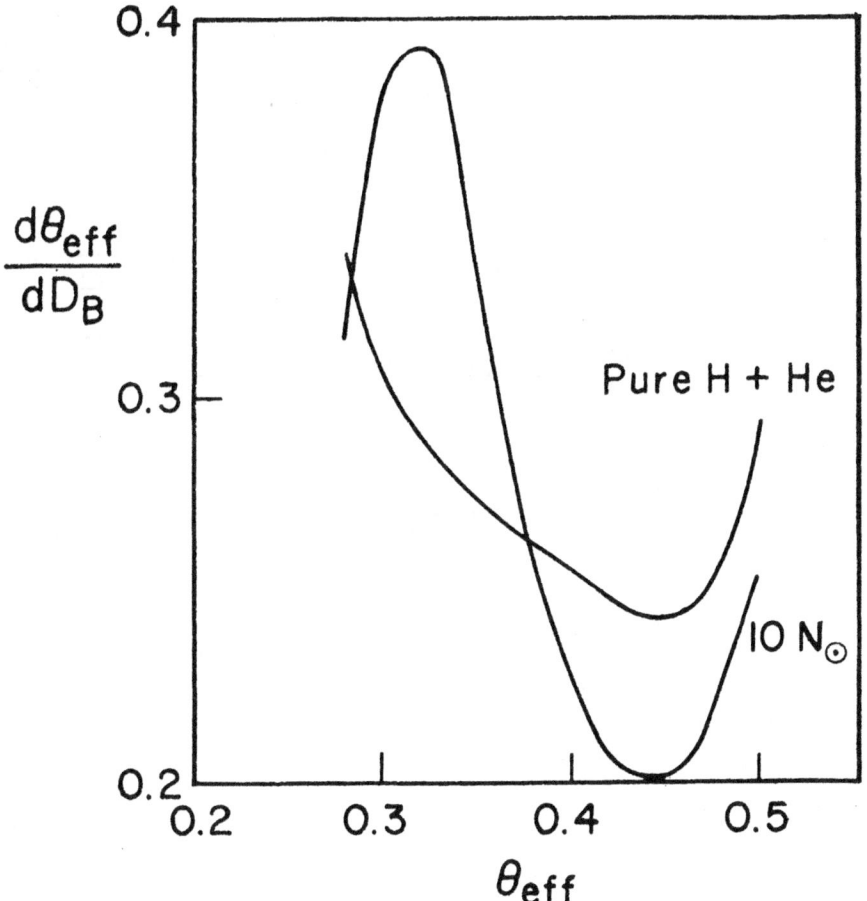

Figure 2. Effect of the nitrogen abundance on the quantity $(d\theta_{eff}/dD_B)$.

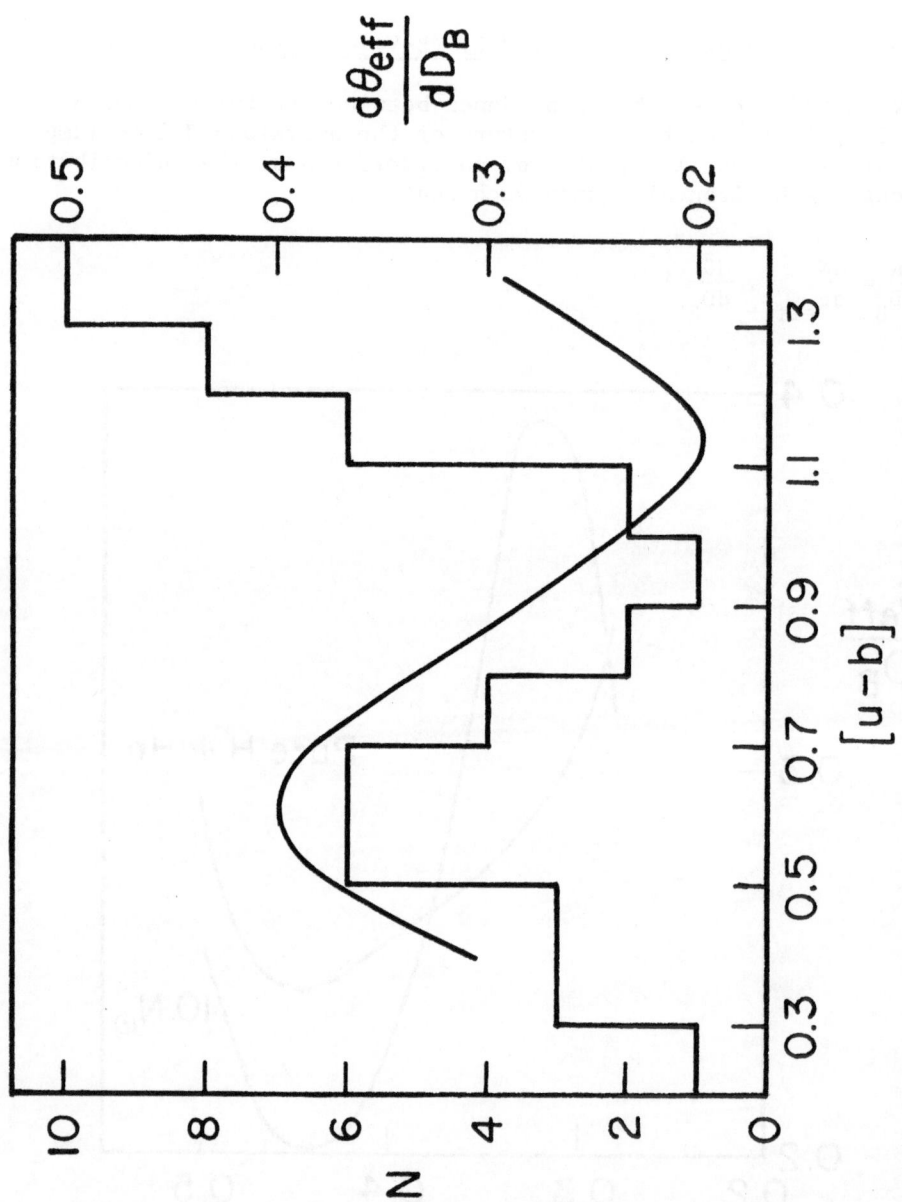

Figure 3. Number distribution of high latitude faint blue stars as a function of temperature superimposed on the curve of Figure 2 for a nitrogen rich atmosphere. The conversion from θ_{eff} to [u−b] is due to Philip and Newell(1975).

If $(dN/d\theta_{eff})$ is a slowly varying function of θ_{eff} in the range of interest, then $N(D_B)$ depends on the form of $(d\theta_{eff}/dD_B)$ in this temperature range. Figure 2, obtained for an atmosphere with ten times the solar nitrogen abundance shows the rapid change in $(d\theta_{eff}/dD_B)$. Since D_B is proportional to $(U-B)$ and θ_{eff} to $(B-V)$, this rapid variation in $N(D_B)$ is in turn reflected as a gap in the $[(U-B)-(B-V)]$-diagram.

The effect is further illustrated in Figure 3 which shows the theoretical curve superimposed on the recent observations of Newell and Graham (1976) of the number distribution of blue halo stars in the field as a function of their [u-b]-color.

In summary, it appears that the variations in D_B on the blue horizontal branch of M92 could be due to large variations in C and/or N at the surface of individual stars. The possible cause of such large effects is still unclear. One cannot rule out at this time the existence of primordial chemical inhomogeneities within the cluster. On the other hand, the nature of the effect suggests the mixing of carbon (with possible subsequent partial or complete processing into nitrogen) from the deep interior as a result of the core helium flash. Already Thomas (1967) has encountered mixing of carbon to the surface in his calculations following the peak of a very energetic flash originating in a shell well removed from the center of the star. Further research on the core helium flash should shed new light on possible mechanisms for mixing in the stellar interior and on the nucleosynthesis which is associated with such mixing.

IV. ACKNOWLEDGMENTS

We are much indebted to Dr. J.B.Oke for pointing out the problem of the M92 Balmer jumps to us and for showing us his unpublished data. This research was supported in part by the National Science Foundation.

REFERENCES

Auer, L.H. and Demarque, P. 1977. Astrophys. J. in press.
Auer, L.H. and Heasley, J.N. Jr. 1973. unpublished.
Böhm-Vitense, E. 1967. Astrophys. J., 150, 483.
Butler, D. 1975. Astrophys. J., 200, 68.
Butler, D., Carbon, D. and Kraft, R.P. 1975. Bull. Am. Astron.Soc.,7,239.
Greenstein, J.L. and Sargent, A.I. 1974. Astrophys.J. Suppl.,28, 157.
Gross, P.G. 1973. Monthly Notices Roy. Astron. Soc.,164, 65.
Johnson, H.L. and Morgan, W.W. 1953. Astrophys.J., 117, 313.
Mihalas, D. 1964. Astrophys.J. Suppl.,9, 321.
Newell, E.B. 1973. Astrophys.J. Suppl.,26, 37.
Newell, E.B. and Graham, J.A. 1976. Astrophys.J., 204, 804.
Oke, J.B. 1975. private communication.
Philip, A.G.D. and Newell, E.B. 1975. Dudley Obs. Report No.9, p. 161.

Sweigart, A.V., Mengel, J.G. and Demarque, P. 1974. Astron. Astrophys.,
    30, 13.
Thomas, H.-C. 1967. Z. Astrophys., 67, 420.
Zinn, R. 1973. Astrophys. J., 182, 183.

CARBON AND NITROGEN ABUNDANCES IN GIANT STARS OF THE METAL POOR GLOBULAR CLUSTER M92 -- A PRELIMINARY REPORT

Duane Carbon
Now at Kitt Peak National Observatory

Dennis Butler
Now at Yale University Observatory

Robert P. Kraft
Lick Observatory, Board of Studies in Astronomy and
Astrophysics, University of California, Santa Cruz

James L. Nocar
Lick Observatory, Board of Studies in Astronomy and
Astrophysics, University of California, Santa Cruz

Observational data on the C^{12}/C^{13} and O/H-abundance ratios in red giants with near solar metal abundance have recently been put forward by Lambert, his associates and students. In summarizing these results, Dearborn, Eggleton, and Schramm (1976) concluded that in the atmospheres of pre-He-core flash red giants there is evidence for substantial depletion of C^{12} and enhancement of C^{13} presumably as a result of deep mixing of envelope material into layers that have undergone CNO-cycle processing. Moreover, values of the ratio C^{12}/C^{13} in the range 10-20 seem to be achieved by stars with luminosities too low, and evolutionary states too early, for compatibility with convectively mixed standard stellar models. It is true, however, as recognized by Dearborn, et al., that red giants of the general star field have checkered backgrounds: giants in the same part of the HR diagram may have quite different masses and ages, even for a fixed initial metal-abundance. Moreover, comparison of a given star with theoretical models may be rendered somewhat uncertain by errors in parrallax or other criteria of absolute magnitude.

One supposes that these problems are largely bypassed in the study of stars in clusters, especially globular clusters, since the evolved stars have closely the same mass, accurately known relative absolute magnitudes, and presumably the same initial chemical composition (We ignore for the moment recent evidence for a Ca abundance spread in ω Cen [cf. Freeman and Rodgers 1975]). In particular, we examine here the carbon and nitrogen abundances in the atmospheres of

Jean Audouze (ed.), CNO Isotopes in Astrophysics, 33-37. All Rights Reserved.
Copyright © 1977 by D. Reidel Publishing Company, Dordrecht-Holland.

giants in the metal-poor globular cluster M92. Earlier work by Zinn
(1973) established that the strength of the G-band of CH depended on
evolutionary state. Thus from examination of small-cycle image tube
spectrograms of 20 giants in M92, Zinn found six stars with unusually
weak G-bands; all were confined to the asymptotic giant branch (AGB).
If present evolutionary ideas are correct (cf. Iben 1974), the six
weak G-band stars belong to a later evolutionary stage than those with
strong G-bands; the latter are confined largely to the subgiant branch
(SGB). The differences are inexplicable on the basis of expected
changes in surface gravity (~0.5 dex) and temperature (very small,
i.e. $\lesssim 100°K$).

A possible explanation for the "Zinn-effect" arises from an
examination of the effective temperatures and surface gravities of the
field halo giants studied by Sneden (1974). For these seven stars,
$\langle [Fe/H] \rangle = -2.2 \pm 0.2$ (m.e.), identical to the metal content of M92 stars
(cf. Butler 1975, and references therein). When these stars are
plotted on the color-magnitude diagram of M92, transforming their
spectroscopically determined values of L and log T_e to M_v and (B-V)
using the model atmospheres of Bell, Eriksson, Gustaffsson and Nordlund
(1976), the stars of highest surface gravity (log g = 2.6 and 2.8) lie
nearest the SGB and M92. According to Sneden, these objects have
$[\frac{C}{H}] \cong [\frac{N}{H}] \cong [\frac{Fe}{H}]$. The other five giants have carbon deficient and
nitrogen overabundant, presumably as a result of the mixing of CNO-
processed material into the external convection zone (Sneden 1974).
These stars have lower surface gravities, therefore lie higher in the
color-magnitude diagram, and thus are compatible with the view that
they are representative of a later evolutionary stage than the stars
with normal C and N.

The combination of Zinn's and Sneden's results leads one to ask
whether the extreme weakness of the G-band in AGB stars of M92 is a
result of the conversion of C into N. The sequence of events supposes
that material just ahead of the H-burning shell, rich in material
processed to equilibrium ratios of $C^{12}:C^{13}:N^{14}$, is somehow mixed into
the convective envelope of an M92 giant, at some moment between the
time of its residence on the SGB and the AGB.

We give a preliminary report here of a test of this proposal based
on observations of both the G-band (CH) and the $\Delta v = 0$ band sequence of
NH near $\lambda 3360$. Scans of 45 giant stars of M92, selected from the
photometric lists of Sandage and Walker (1966) and Cathey (1974), were
obtained with the image-dissector scanner (IDS) (Robinson and Wampler
1972), operated at the Cassegrain focus of the Lick 120-inch (3 m)
reflector. The scanner output, fed into the Miller spectrograph, was
used in conjunction with a 600 line/mm grating blazed at λ = 5000 A in
the 1st order, which produced an output dispersion with the blue-sensi-
tive image tube chain near 120 A/mm. The spectral interval covered by
the 2048 channels of the scan was roughly $\lambda\lambda 3100$-5200A. All spectra
were normalized by the observation of blue standard stars of known
energy distribution (Hayes 1970, Stone 1974), and fluxes F_λ were

derived across the entire spectral range. Six of the seven field
giants studied by Sneden (1974) were also observed with the same equip-
ment and were used as control over the method of analysis. Further
observational details and an extensive description of the photometric
accuracy will be given in a forthcoming paper to be published in the
Astrophysical Journal.

Our method of analysis consists of comparison of the observed
flux-corrected scans with synthetic spectra computed from model atmospheres for metal-poor giants (Bell, et al. 1976). Synthetic
reproduction of several hundred angstroms of spectrum in the intervals
$\lambda\lambda 3200-3700$ A and $\lambda\lambda 4250-4450$ enables us to estimate values of [Fe/H],
$[\frac{C}{H}]$, and $[\frac{N}{H}]$ good to a factor of about 3. Further adjustments of the
f-values of individual lines, derived in all cases by reproducing
corresponding portions of the Kitt Peak Solar Atlas, are currently in
progress; we shall therefore not attempt at this time to give final
estimates for [C/H] and [N/H] but rather values believed correct to
first order. It is well to note that many M92 giants are quite faint
($m_{pg} \gtrsim 16$) and therefore the spectra obtained even with the ITS are of
small scale and resolution. Thus although Fe, C, and N abundances can
be estimated with some accuracy, we give up all hope of obtaining
isotope ratios such as C^{12}/C^{13}.

Statistically satisfactory scans have been obtained for 45 giants
in M92: 20 stars belonging to the SGB, 6 to the AGB, and 19 to the
ordinary giant branch (GB), the last defined to consist of those stars
with (B-V) > 0.80 and M_v brighter than -0.1. We summarize our results
as follows, turning our attention in the beginning to carbon. First,
the "Zinn effect" is confirmed, i.e., stars on the AGB have very weak
G-bands in comparison with stars of the same (B-V) on the SGB. We
estimate that carbon abundances are normal for most SGB stars, i.e.,
$[\frac{C}{H}] = [\frac{Fe}{H}] = -2.2$, but AGB stars in general have $[\frac{C}{H}] \sim -3.0$. Second,
stars of the ordinary giant branch show a marked deficiency of carbon
relative to stars of the SGB. Eleven of these stars show carbon
reduced by a factor of 2 to 5, whereas the balance of eight show carbon
reduced by a factor $\gtrsim 5$. Third, the changeover from normal carbon to a
moderate reduction in carbon (factor of 2 to 5) occurs high on the SGB
at an average location near $M_v \sim +1.0$. Above this level, there are
three stars with normal carbon; below, there are three stars with
slightly reduced carbon. The changeover is therefore not an extremely
sharp function of luminosity, but nevertheless definitely occurs on the
SGB. There is no evidence that, at a fixed luminosity on the SGB, the
carbon abundance is related to the observed "spread" in (B-V).

The material at hand suggests that the products of CNO processing
are convected into the atmosphere prior to the helium core flash when
the star reaches absolute magnitude $M_v \sim +1.0$, and that the atmospheric
carbon abundance is lowered by a factor of 2 to 5 which persists all
the way to the red-giant tip. Further reduction appears to have taken
place prior to the second ascent of the giant branch along the AGB.
Although other explanations are possible, in this tentative picture

the eight stars of the GB, those having carbon greatly reduced to a
value like that of the AGB stars, are then in the second ascent to the
red giant tip. Such a large carbon reduction may be understandable as
a result of the high temperature reached at the base of the convection
zone in helium shell source stars (cf. Scalo, et al. 1975). It is also
of interest that the observed absolute magnitude ($M_v \sim +1.0$) of the SGB
changeover point is not far from the absolute magnitude ($M_v \sim +1.5$) at
which one-solar mass metal-rich models, of the kind considered by
Dearborn, et al. (1976), have their deepest convective mixing.

The above discussion suggests a carbon behavior that is reasonably
orderly and understandable as a result of convective mixing and CNO
processing; however, it is beyond the scope of these remarks to consider
whether the numerical values of the abundance estimates are compatible
with mixed model calculations. We turn instead to the results for
nitrogen. In contrast with the situation for carbon, the nitrogen
abundances are relatively chaotic and difficult to understand. Although
a few stars of the SGB have normal nitrogen, i.e., $[N/H] = [\frac{C}{H}] = [\frac{Fe}{H}]$,
a larger fraction have what appear to be large nitrogen enhancements.
SGB stars having virtually identical (B-V)'s and M_v's, can have extra-
ordinarily different spectral appearance at $\lambda 3360$ (NH), yet their
G-bands can be quite comparable. In addition, stars of the AGB,
although having as a class much reduced G-band strengths, do not have
an enhancement of the $\lambda 3360$ feature beyond that exhibited by a large
fraction of stars of the SGB. In short, the expected anti-correlation
between the strengths of features attributed largely to CH and NH does
not show up in any simply-interpretable way.

The meaning of these results is presently quite unclear. In
contrast to the ω Cen situation, we find no evidence for variations in
Fe-abundance in M92 stars, so the enhancement of N is not accompanied,
for example, by a corresponding enhancement of Fe. The existence of
stars with large N enhancements lying side-by-side on the SGB with
stars having normal N is not immediately explicable by arguments based
on stellar evolution. We are presently investigating cluster models
with multiple stellar populations; in any case, it is hard to understand
how nitrogen can be strongly enhanced in stellar systems of very old age.

We are indebted to Drs. R. Bell and B. Gustafsson for communication of their model atmosphere results in advance of publication.

REFERENCES

Bell, R., Eriksson, K., Gustafsson, B., and Nordlund, A. 1976,
A. and A. Supple., 23, 37.

Butler, D. 1975, Ap. J., 200, 68.

Cathey, L. 1974, A. J., 79, 1370.

Dearborn, D., Eggleton, P. and Schramm, D. 1976, Ap. J., 203, 455.

Freeman, K. and Rodgers, A. 1975, Ap. J. Letters, 201, L71.

Hayes, D. 1970, Ap. J., 159, 165.

Iben, I. 1974, Ann. Rev. Astron. and Ap., 12, 215.

Robinson, L. and Wampler, E. 1972, Pub. A. S. P., 84, 161.

Sandage, A. and Walker, M. 1966, Ap. J., 143, 313.

Scalo, J., Despain, D. and Ulrich, R. 1975, preprint.

Sneden, C. 1974, Ap. J., 189, 493.

Stone, R. 1974, Ap. J., 193, 135.

Zinn, R. 1973, Ap. J., 182, 183.

CNO ISOTOPES AND RED GIANTS

D. S. P. Dearborn
Institute of Astronomy, Madingley Road, Cambridge, England.

I am going to talk about the use of CNO observations as an indicator of phenomena which are important to a star's evolution. During the cool red giant stage, molecular lines are observed from which CNO isotope ratios and abundances can be determined. It is also during this stage that a surface convection zone penetrates into the interior mixing out material which was processed via the CNO tricycle (Iben 1964, 1966). Comparison of observed isotope ratios (or abundances) with theoretical expectations can then be used as a check on the occurrence of such phenomena as convection, meridional mixing, and mass loss.

Isotope ratios can be more easily and accurately determined than absolute abundances because they are less model dependent. The relatively low ratio of ^{12}C to ^{13}C makes it the easiest of the CNO isotope ratios to observe. Calculations by Dearborn, Schramm and Eggleton (1976, = DSE) show that the expected $^{12}C/^{13}C$ ratio ranges from 20 to 30. When compared to the observed values in Table I it is seen that some stars agree with theoretical expectations while others show significant enhancement of ^{13}C. In particular, α Boo and γ Leo A are spectacularly discrepant, and these stars have too low a luminosity for the low $^{12}C/^{13}C$ ratio to be attributed to such complications as double shell flashes (Iben 1975, Gingold 1974) or hot bottomed convection zones (Scalo et al 1975). In DSE a number of possible explanations of the low $^{12}C/^{13}C$ ratios were explored including:
    a) variations in the initial $^{12}C/^{13}C$ ratio
    b) meridional mixing
    c) mixing at the helium core flash
    d) mass loss prior to deep mixing.

To explain the low $^{12}C/^{13}C$ ratios by initial variation requires inhomogenuity in the interstellar medium too large to be consistent with observations (Wannier et al 1976), and mixing at core helium flash seems more likely to enhance ^{12}C than ^{13}C. Therefore DSE concluded that either meridional mixing on mass loss prior to the deep mixing seemed the most likely causes of low $^{12}C/^{13}C$ ratios.

Jean Audouze (ed.), CNO Isotopes in Astrophysics, 39-44. All Rights Reserved.
Copyright © 1977 by D. Reidel Publishing Company, Dordrecht-Holland

Table I

	Star		$^{12}C/^{13}C$	reference
(a)	δ	Eri	>50	1
	η	Cep	>30	1
	ν²	CMa	51	1
(b)	γ	Cep	24	1
	i	Cep	16	1
	λ	Sgr	22	1
	δ	Tau	23	5
	ρ	Boo	15	1
	δ	And	25	1
	σ	Boo	7.2	4
	γ	Leo A	6.5	2
	α	Tau	9	2
(c)	ξ	Cyg	22	2
	ε	Peg	5.1	3
	ψ	And	27	5
	ε	Gem	6	5
	α	Ori	7	6

Stars in group (a) have luminosities low enough (log L<1.0) to be consistent with not having mixed the ^{13}C to the surface, or to have just begun such mixing. Group (b) stars (1.0<log L<2.75) should have undergone only deep convective mixing on the giant branch. Group (c) stars (log L>2.75) may have been exposed to double shell flashes which could affect the $^{12}C/^{13}C$ ratio. The references for these are:

1 Dearborn et al 1975   4 Day et al 1974
2 Tomkin et al 1975   5 Tomkin et al 1976
3 Lambert and Tomkin 1974   6 Lambert et al 1974.

    Recently Dearborn and Eggleton (1976) studied the effect on the CNO isotopes of deep mixing while on the main sequence. The calculations used a diffusion approximation which although one dimensional should represent the effect of meridional mixing (Zahn 1975). The two important parameters in such mixing are the timescale of mixing (τ_{mix}) and the depth to which the mixing operates (M_R). Figure 1 shows the results for a 3.0 $M_\odot$ star, and is representative of stars in the range 1.5 to 6 $M_\odot$. Stars of higher mass are not significantly affected by meridional mixing (Paczynski 1973) and the steeper gradient of mean molecular weight (due to the increased importance of the pp chain) in lower mass stars inhibits such mixing.

    Figure 1 can be considered in two sections; fast mixing where the mixing timescale is less than the main sequence timescale ($\tau_{mix} < \tau_{ms}$), and slow mixing where the mixing timescale is longer than the main sequence lifetime ($\tau_{mix} > \tau_{ms}$). Fast mixing is very sensitive to the

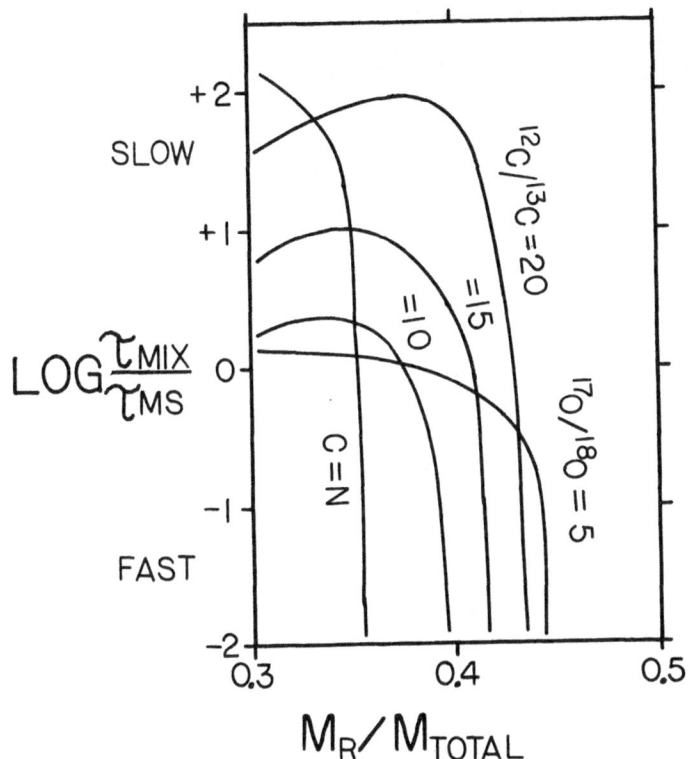

Figure 1.   The effects of diffusion mixing on CNO isotopes are shown for different mixing timescales (τ_{mix}) and depths of mixing (M_R). Contours are given for $^{12}C = ^{14}N$, $^{17}O/^{18}O = 5$, and $^{12}C/^{13}C = 10, 15, 20$.

depth of mixing. Mixing down to $M_R/M_{total} = 0.43$ has almost no effect on the $^{12}C/^{13}C$ ratio of a 3 $M_\odot$ star, while mixing slightly deeper results in a very low $^{12}C/^{13}C$ ratio. Rapid mixing also depletes ^{18}O and so causes a large $^{17}O/^{18}O$ ratio. Most low $^{12}C/^{13}C$ ratios lie in the range of 10 to 20. In order to produce values in this range, fast mixing must operate primarily to a very narrow range of depth. This is diffi-cult to understand in stars of 2 to 6 $M_\odot$ where the gradient of the mean molecular weight is very shallow, and so mixing should proceed completely through this region. One should expect fast mixing to produce many stars with very low $^{12}C/^{13}C$ ratios (<6).

Slow mixing is insensitive to depth, depending instead on the mixing timescale. ^{18}O is not greatly affected, but enhancement of ^{13}C usually requires a large enhancement of ^{14}N. If stars with low $^{12}C/^{13}C$ are shown to have enhanced ^{14}N such that $^{14}N > ^{12}C$, slow meridional mixing would seem a likely cause. That meridional mixing is slow also agrees with the suggestion by Schwarzschild (1947) that it is inefficient at momentum (and material) transport.

As mentioned above, mass loss prior to the onset of deep convective mixing can affect observed isotope ratios. OB stars are observed to have large mass loss rates on the order of 10^{-7} to 10^{-5} $M_\odot$/years (Morton, 1967, Hutchings 1970, Hearn, 1975). Dearborn and Eggleton (1976) have calculated the effect of such mass loss rates on massive stars. It was found that mass loss rates $M > 1.5 \times 10^{-6}$ $M_\odot$/years removed sufficient mass from a 32 M star to expose material processed via the CN cycle. Such stars would appear to be OBN stars. Larger mass loss rates reduce the luminosity somewhat, but produce OBN stars near the zero age main sequence. When these stars evolve to become red giants, they have very low $^{12}C/^{13}C$ ratios comparable with the values observed in α Ori, ε Peg and ε Gem.

Other isotope ratios have been observed (Rank, Geballe and Wollman 1974, Maillard 1973). Observations of the oxygen isotope ratios can offer much additional information. As was mentioned earlier, a high $^{17}O/^{18}O$ ratio (>5) could be an indication of fast meridional mixing in stars of low $^{12}C/^{13}C$. The $^{17}O/^{18}O$ ratio could also be useful in determining the maximum depth that the surface convection zone penetrates into a star. Since the CN cycle converts ^{12}C primarily into ^{14}N, mixing of this material to the surface has little effect on the $^{12}C/^{13}C$ ratio. The primary effect on $^{12}C/^{13}C$ comes from mixing a thin region rich in ^{13}C near $M_R/M_{total} \sim 0.4$. The $^{12}C/^{13}C$ ratio is therefore insensitive to the exact depth of mixing. The ^{17}O abundance however is low in the envelope and increases rapidly towards the core. The final $^{17}O/^{18}O$ ratio depends sensitively on the depth to which the surface convection zone penetrates. Figure 2 shows the resulting $^{17}O/^{18}O$ ratios for different depth of mixing in 1.5 and 2 $M_\odot$ stars.

In the end, all stars give up the gas and effect the interstellar medium. It is possible to calculate the expected ^{13}C enhancement due to envelope ejection by red giants. This can be compared to the expected ^{12}C enhancement from supernovae to determine the evolution of the $^{12}C/^{13}C$ ratio. Pioneering calculations of this sort were done by Wollman (1973) and Audouze et al (1975). As a better understanding is obtained of the phenomena effecting stars, the uncertainties in such calculations will decrease. Ultimately such studies aid in understanding the chemical evolution of the galaxy.

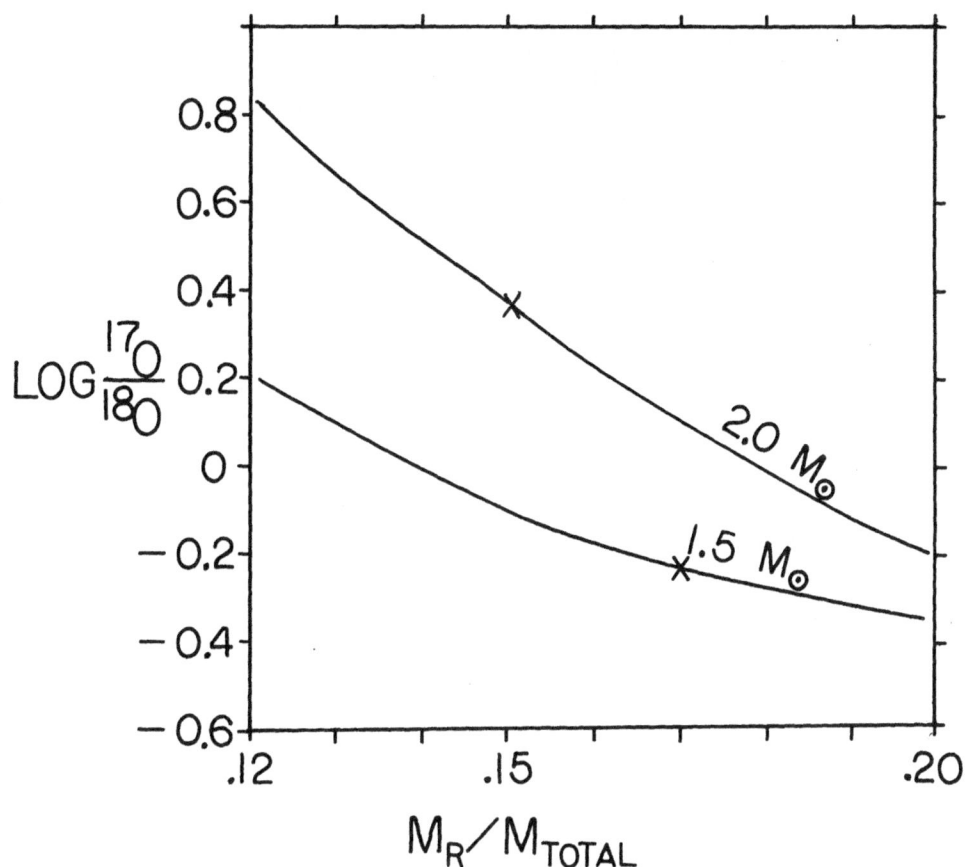

Figure 2. The expected $^{17}O/^{18}O$ ratio for simple deep mixing as a function of depth of mixing. The symbol shows the maximum depth of convective penetration expected from mixing length theory with $\alpha = 1.5$.

References

Audouze, J., Lequeux, J. and Vigroux, L., 1975, Astr. and Astrophys., 43, 71.
Day, R., Lambert, D. and Sneden, C., 1974, Astrophys. J., 185, 213.
Dearborn, D., Lambert, D. and Tomkin, J., 1975, Astrophys. J., 200, 675.
Dearborn, D., Schramm, D. and Eggleton, P.,1976,Astrophys.J., 203, 455.
Dearborn, D. and Eggleton, P., 1976, Astrophys. J., in press.
Dearborn, D. and Eggleton, P., 1976, Astrophys. J., in press.
Gingold, R., 1974, Astrophys. J., 157, 737.
Hearn, A., 1975, Astr. and Astrophys., 40, 277.
Hutchings, J., 1970, MNRAS, 147, 161.
Iben, I., 1964, Astrophys. J., 140, 1631.
Iben, I., 1966, Astrophys. J., 143, 483.

Iben, I., 1975, Astrophys. J., 196, 525.
Lambert, D. and Tomkin, J., 1974, Astrophys. J. (Letters), 194, L89.
Lambert, D., Dearborn, D. and Sneden, C., Astrophys. J., 193, 621.
Maillard, J., 1973, Highlights of Astronomy, 3, z69.
Morton, D., 1967, Astrophys. J., 147, 1017.
Pacynski, B., 1973, Acta Astr., 23, 191.
Rank, D., Geballe, T. and Wollman, E., 1974, Astrophys. J. Letters, 187,
        L111.
Scalo, J., Despain, D. and Ulrich, R., 1975, Astrophys. J., 196, 805.
Schwarzschild, M., 1947, Astrophys. J., 106, 407.
Tomkin, J., Lambert, D. and Luck, E., 1975, Astrophys. J., 199, 436.
Tomkin, J., Luck, E. and Lambert, D., 1976, preprint.
Wannier, P., Penzias, A., Linke, R. and Wilson, R., 1976, Astrophys. J.,
        209, 26.
Wollman, E., 1973, Astrophys. J., 184, 773.
Zahn, J., 1975, Mem. Soc. Roy. Sci. Liége, Tome VIII, 31.

ABUNDANCE OF CARBON IN NGC 7027

N. PANAGIA
Laboratorio di Astrofisica Spaziale, Frascati and
Laboratorio di Radioastronomie, Bologna, Italy

E. BUSSOLETTI, A. BLANCO
Istituto di Fisica, Università di Lecce, Italy

In a recent study of the spectrum of the planetary nebula NGC 7027 (Panagia et al. 1976) we were able to determine both the amount of carbon contained in grains and the amount in a gaseous form. Here we present the most relevant results and discuss some implications on the late stages of the stellar evolution and the chemical evolution of the Galaxy.

1. CARBON IN GRAINS

We have examined the problem of energy balance of the dust present within the ionized region, i.e. the dust emission must balance the energy gained by absorption of stellar and nebular radiation. We have computed a series of models in which the main paramter is the optical depth of dust in the Lyman continuum region (τ_{uv}). Then, by equating the total infrared flux, derived from the observations, to the power absorbed by the dust, as obtained from our models, we have found $\tau_{uv} = 0.29 \pm 30$ %. Also, the attenuation of the CIV line at 1550 A due to internal dust has been found to amount to a factor 3.8 ± 0.8.

The dust has been identified with graphite because it is the only plausible material which can fulfill the following requirements : 1) to have an high abundance as implied by the infrared luminosity (see below); 2) to have a smooth, featureless absorptivity as implied by the IR spectrum (Forrest, private communication) ; 3) to have an high sublimation temperature so that it can condense ar relatively high temperatures and survive in the intense radiation field of a proto-planetary nebula ; 4) to condense from the same medium and at the same conditions as for the condensation of carbonates (which are responsible for the 11.2 μm feature, Bregman and Rank, 1975).

From an anlysis of the infrared spectrum we have found that a range of grain temperatures is needed, with $T_{min} \simeq 90$ K and $T_{max} \simeq 600$ K. This range can only be accounted for by a grain size distribution with a maximum grain radius $a_{max} \simeq 1.4$ μm and a minimum one a_{min} 1.2×10^{-3} m.

Once the optical depth, the nature and the size distribution of the dust are known, the mass of the dust present in the ionized region can be derived as $M(dust) = 4.8 \times 10^{-4} D^2 \pm 40 \% M_\odot$, D being the distance in Kpc. The mass of the ionized hydrogen has been estimated from the radio flux to be $M(HII) = 4.3 \times 10^{-2} D^2 \pm 30 \% M_\odot$. Therefore, the abundance by number of carbon contained in grains relative to gaseous hydrogen is $C(dust)/H = 9.4 \times 10^{-4}$ with an indetermination of a factor 1.57.

2. ABUNDANCE OF GASEOUS CARBON

From the measurements of Bohlin et al. (1975) the intensities of the ultraviolet emission lines of CIII at $\lambda = 1909$ Å and CIV at $\lambda = 1550$Å, corrected for both internal and interstellar absorption, are $I(1909) = 12.81(H\beta)$ and $I(1550) = 64.61(H\beta)$, respectively. The contribution of recombinations (radiative + dielectronic) to the line intensities has been found to be negligible. Thus, electronic collisions are chiefly responsible for the formation of these lines. Furthermore, in both cases the lifetime against radiative decay is extremely short, so that collisional de-excitation is quite negligible and the derivation of the abundances should be straightforward. However, a knowledge of the electron temperature is needed. Very different estimates of T_e for NGC 7027 can be found in the literature : they range from 1.1×10^4 K (Kaler et al. 1976) up to 1.9×10^4 K (Churchwell et al. 1976). Using the results of a very recent analysis of the CII λ 4267 Å line (Peimbert and Torres-Peimbert, 1976) we have estimated that $T_e > 1.6 \times 10^4$ K in the zone of existence of C^{++}. Since the highest possible temperature is 1.9×10^4 K, we have adopted $T_e(C^{++}) = (1.75 \pm .14) \times 10^4$ K. By also adopting this temperature for C^{3+}, the result is $C^{++}/H^+ = 3.6 \times 10^{-4}$ and $C^{3+}/H^+ = 2.5 \times 10^{-4}$ by number, with a possible indetermination of a factor 1.56. Allowing for four times ionized carbon, the total abundance of gaseous carbon relative to hydrogen becomes $C(gaseous)/H = 9.4 \times 10^{-4}$ by number.

3. DISCUSSION

Summarizing the above results, the total abundance of carbon in NGC 7027 is $C/H = 1.9 \times 10^{-3}$ by number ($C/H = 2.2 \times 10^{-2}$ by mass), with a possible determination of a factor 1.6. About half of it is contained in grains and half is in a gaseous form. The total abundance of carbon is about 5.7 times higher than the " cosmic " one (cf. Allen 1973).

It is instructive to compare this with the abundance of other relevant elements, namely He, N, O. The abundance of He is $He/H = 0.126$ by number (Peimbert and Torres-Peimbert, 1971 ; Miller and Mathews, 1972) with little uncertainty (< 15 %). The abundances of N and O instead are less accurately known, because the physical conditions within the nebula (n_e and T_e) are not well determined. Therefore, we adopt $N/H = 2.2 \times 10^{-4}$ and $O/H = 5.7 \times 10^{-4}$ which are straight averages of the values found in the literature. The estimated indetermination is a factor 1.6 for both elements. Despite the large uncertainties, it is

evident that, besides C, He and N are also overabundant relative to
cosmic abundances (by factors of 1.46 and 2.2, respectively), whereas
O is normal or perhaps slightly underabundant. (However, part of the
oxygen could be contained in carbonate grains ; the maximum deficiency
of oxygen would be about $\Delta O/H \simeq 3(Mg/H + Fe/H) \simeq 2 \times 10^{-4}$ by number).
Thus, there is clear evidence that some contamination of the original
material by both the products of H-burning, through the CNO-cycle, and
He-burning has occurred. By comparing the actual abundances of NGC 7027
to the solar ones, we estimate that about 10 % of the nebular matter
has undergone some nuclear processing. Furthermore, since oxygen is not
overabundant while carbon is, the matter which has undergone He-burning
must have been processed either incompletely or at a relatively low
temperature in order to have produced little oxygen relative to carbon.
It is interesting that the calculations of Iben (1976) predict that the
only products of He-burning, which can be brought to the surface of a
star without an explosive events, come from a He-burning shell in which
only a fraction ($\sim$ 0.2) of the helium mass has been converted into carbon
and in which practically no oxygen has been produced. Moreover, the
fact that carbon is more overabundant than nitrogen indicates that
carbon, after being produced via 3α-reaction, has not been substantially
processed via CNO-cycle in a H-burning zone. This indication finds con-
firmation in the fact that the isotopic ratio $^{12}C/^{13}C$ is higher than,
or at least equal to 20, which is the ratio of the observed line intensities of ^{12}CO to ^{13}CO in the molecular cloud associated with NGC 7027
(Mufson et al. 1975).

From the above results we infer that the parent star of NGC 7027
was a carbon star when it was in the last red giant phase. The fact
that C and N are more abundant in NGC 7027 than in even the coolest
carbon stars (cf. Kilston, 1975) may indicate that the last episode(s)
of mixing and mass loss, which produced the planetary nebula envelope,
has concerned deeper layers of the star than the previous episodes.
Within this frame, it is interesting to note that the spectra of the
central stars of the planetary nebulae usually exhibit stronger carbon
lines relative to other lines than in " normal " early type stars of
the same spectral type (cf. Smith and Aller, 1969). Thus, we speculate
that most, if not all, planetary nebulae are strongly overabundant of
C and moderately overabundant of N and He.

Finally, we estimate what the contribution of planetary nebulae can
be to the chemical evolution of the Galaxy. For the return of mass from
planetary nebulae to the interstellar medium, we take a value of 0.2
$M_\odot/y$ (Alloin et al. 1976). This may be an underestimate because it corresponds to the assumption that the mass lost by the parent red giant
to form a planetary nebula is 0.16 $M_\odot$, whereas it could be significantly
higher (Fusi-Pecci and Renzini, 1976). Also, we assume the abundances
of NGC 7027 to be typical of all planetary nebulae : this may be an
overgenerous assumption, but hopefully it compensates for an underestimate of the ejected mass. Then, assuming that the production rate of
planetary nebulae in the Galaxy has been the same in the past as it is
now, we find that planetary nebulae contribute significantly to the

carbon enrichment of the Galaxy, i.e. ΔC(from planetary nebulae)/C(present) $\sim$ 0.1. This fraction could be even higher if the rate of production of planetary nebulae had been higher in the past. However, the average rate in the whole lifetime of the Galaxy cannot be higher than ten times the present one, otherwise one would produce more carbon than that which is observed. With this limitation, we estimate that the enrichment of the Galaxy in He and N by planetaries is not as important as for C, ranging from $\Delta He/He = 8.7 \times 10^{-3}$ and $\Delta N/N = 4.2 \times 10^{-2}$ (present rate of production of planetary nebulae) up to the maximum values $\Delta He/He = 8.7 \times 10^{-2}$ and $\Delta N/N = 0.42$ (10 x present rate).

E. Bussoletti and A. Blanco were supported in part by NATO grant N° 861.

REFERENCES

Allen, C.W., 1973, " Astrophysical Quantities ", The Athlone Press
    (London), p. 21.
Alloin, D., Cruz-Gonzales, C. and Peimbert, M., 1976, Astrophys. J.,
    205, 74.
Bohlin, R.C., Marioni, P.A. and Stecher, T.P., 1975, Astrophys. J.,
    202, 415.
Bregman, J.D. and Rank, D.M., 1975, Astrophys. J. (Letters) 195, L125.
Churchwell, E., Terzian, Y. and Walmsley, M., 1976, Astron. & Astrophys.
    48, 331.
Fusi-Pecci, F. and Renzini, A., 1976, Astron. & Astrophys., 46, 447.
Kaler, J.B., Aller, L.H., Czyzak, S.J. and Epps, H.W., 1976, Astrophys.
    J. Suppl. Ser. 31, 163.
Kilston, S., 1975, Pub. A.S.P., 87, 189.
Iben, I., 1976, Astrophys. J., 208, 165.
Miller, J.S. and Mathews, W.G., 1972, Astrophys. J., 172, 593.
Mufson, S.L., Lyon, J. and Marionni, P.A., 1975, Astrophys. J. (Letters)
    201, L85.
Panagia, N., Bussoletti, E. and Blanco, A., 1976, Astron. & Astrophys.,
    Submitted for publication.
Peimbert, M. and Torres-Peimbert, S., 1971, Astrophys. J., 168, 413.
Peimbert, M. and Torres-Peimbert, S., 1976, in preparation.
Smith, L.F. and Aller, L.H., 1969, Astrophys. J., 157, 1245.

CNO NUCLEOSYNTHESIS AND THE NOVA OUTBURST

Sumner Starrfield
Department of Physics
Arizona State University, Tempe, Arizona

James Truran
University of Illinois Observatory
Urbana, Illinois

Warren Sparks
Goddard Space Flight Center
Greenbelt, Maryland

I. INTRODUCTION

In this review we will present our predictions for CNO nucleosynthesis by the classical nova outburst. Inasmuch as a detailed review of the outburst itself will soon appear in print (Starrfield, Sparks, and Truran, 1976, hereafter SST), here we will discuss only those properties of the nova phenomena which are pertinent to the production of the CNO isotopes. We will also discuss some observational studies of CNO abundances in novae which strongly support our theory for the cause of the outburst.

It is now commonly assumed that all novae are close binaries with one star filling its zero-velocity lobe and the other star a small hot white dwarf. The lobe filling star is in a stage of evolution where it is slowly losing mass through the inner Lagrangian point into the lobe surrounding the white dwarf. This gas spirals into an accretion disc and some fraction reaches the white dwarf's surface. As the accretion process continues, a layer of hydrogen-rich material will be built up on the white dwarf until the bottom of the hydrogen is compressed and heated to nuclear burning temperatures. A thermonuclear runaway ensues and if the CNO nuclei are enhanced in the envelope this process produces a fast nova outburst. If the CNO nuclei are not enhanced, this process produces a slow nova outburst.

We know nothing about the abundances in the transferred material and in all our studies have assumed that it had solar abundances. In fact, the observations of dwarf novae and accretion discs in old

Jean Audouze (ed.), CNO Isotopes in Astrophysics, 49-61. All Rights Reserved.
Copyright © 1977 by D. Reidel Publishing Company, Dordrecht-Holland.

novae show no obvious abundance anomalies (Warner, 1976). On the
other hand, we would not be surprised to find abundance anomalies in
this gas since it seems unlikely that we are observing cataclysmic
variables in the first phase of mass transfer. The earlier stage of
mass transfer, from the present white dwarf to the present lobe filling star, could have transferred nuclear processed material. Abundance anomalies would also be expected if cataclysmic variables are
the descendants of Case C mass exchange (Ritter, 1976).

We use the hydrostatic studies of accretion onto a white dwarf
(Giannone and Weigert, 1967; Taam and Faulkner, 1975; Colvin, et.al.,
1977) to justify our initial models that have the hydrogen in place
and in equilibrium. These studies show that a white dwarf can accrete
a layer of hydrogen and that at least $10^{-4}M_\odot$ of material must be accreted by $1.00M_\odot$ white dwarfs before the runaway can begin. They also
show that the thickness of the layer depends on the mass and luminosity
of the white dwarf. We follow the evolution with a lagrangian, one-
dimensional, hydrodynamic stellar evolution computer code which is
fully implicit in both the thermodynamics and kinetics (Kutter and
Sparks, 1972) and obtain the nuclear energy generation from a reaction
network which follows the abundances of 10 CNO nuclei plus hydrogen
and ^{4}He (Starrfield, et.al., 1972). Recently, we have extended our
network for the proton-proton chain to include ^{7}Be and ^{8}B in order to
determine the abundance of ^{7}Li in the ejected material. The importance of the CNO network is that it allows us to treat the β^+-unstable
nuclei correctly and, therefore, determine the cause of the classical
nova outburst.

Since the bottom of the layer becomes electron degenerate and is
fairly thick, one can expect the energy released from the nuclear reactions to drive the temperature to 10^8 °K and higher. At these temperatures, the proton capture rates become faster than the β^+-decay
half lives of ^{13}N, ^{14}O, ^{15}O, and ^{17}F completely changing the behavior
of the energy generation.

As was presented in SST, the result of this process was an explosion on the white dwarf that resembled the features of the Nova Outburst: at least $10^{-5}M_\odot$ of material was ejected; with ejection speeds
from 200 to 3000 km/sec, and kinetic energies in the ejecta of 10^{44} to
10^{45} ergs. However, SST also predicted, on the basis of non-equilibrium nuclear burning, that the CNO isotopes in the nebula would have
extremely non-solar abundances and isotopic ratios.

In the following section we shall discuss the effect of the β^+-
unstable nuclei on the evolution, the need for enhanced CNO nuclei in
the envelope, and possible mechanisms for producing this enhancement.
The next section will be devoted to the recent observations of enhanced
CNO nuclei in the nova ejecta. The final two sections present the evolutionary results and a summary and discussion of our predictions.

II. THE EFFECT OF THE β^+-UNSTABLE NUCLEI ON THE EVOLUTION

Even though we have continuously emphasized the importance of the β^+-unstable nuclei and their delayed energy production (c.f., SST and references therein), it is still useful to discuss how they effect the evolution and produce the outburst. This section is even more relevant with the discovery that unenhanced models can produce slow-novae like outbursts (Sparks, 1976a, 1976b, Prialnik et.al., 1976).

The importance of the β^+-unstable nuclei arises because of the high temperatures in the shell source ($T>10^8$ $^\circ$K) which will have the following effects: First, because the proton captures occur much more quickly than the β^+-decays, at maximum temperature all of the CNO nuclei in the shell source will have been converted to the β^+-unstable nuclei and the energy generation at this time will come from a β^+-decay followed by a proton capture. This implies that the peak rate of energy generation will depend on the total amount of CNO nuclei initially present in the envelope since the CNO reactions do not produce new CNO nuclei but only re-distribute those that were initially present. Second, since the β^+-decays are neither temperature nor density dependent, peak energy generation will be followed by a phase with a long, slow decline in nuclear burning. Third, the initial rise to peak burning releases enough energy to cause a general expansion of the envelope. This means that the energy release from the β-decays will occur when the envelope is not as tightly bound to the white dwarf, making is significantly easier to eject material during the outburst. Therefore, we find that the energy release from the β^+-decays is responsible for the final use in luminosity of the outburst and the ejection of the nebula. Next, since most of the nuclei decay after the envelope has expanded and become too cool for further proton captures, the final isotopic abundance ratios that we predict will not resemble equilibrium CNO cycle burning at any temperature. However, the actual isotopic ratios will depend on the temperature and density history in the envelope. Finally, one model dependent result is that the growing shell source produces a convective region which reaches to the surface. This will not only carry the β^+-unstable nuclei outward so that the energy released near the edge of the star can exceed 10^{12}erg gm^{-1}sec^{-1}, but also will mix in fresh unburned nuclei during the peak of the flash keeping the nuclear reactions in the shell source far out of equilibrium.

Implicit in this discussion is our discovery that the envelope must be enhanced in the CNO nuclei in order to produce the peak energy generation and total energy production necessary for mass ejection and the observed light curves of the fast nova (Starrfield, Truran, Sparks, Kutter, 1972). We have already presented order-of-magnitude arguments which demonstrate the need for enhanced energy production and, therefore, enhanced abundances (SST) and we will not repeat them here. The most exciting developments of the past year seem not only to indicate where these enhancements might come from but, also, to indicate that enhanced CNO nuclei have been observed in novae ejecta.

We have generally assumed that the outburst occurs on a carbon oxygen white dwarf in a close binary system and that somehow nuclei from the core are mixed into the envelope. There are two major difficulties with this hypothesis: First, it has been commonly assumed that the cataclysmic variables are the descendants of the WUMa systems and it has always seemed very unlikely that these binaries could evolve into a configuration where one star is a massive white dwarf with a carbon/oxygen core. Recently, however, Webbink (1976) has shown that low-mass contact binaries are subject to a dynamic mass-transfer instability and that they cannot be the progenitors of the cataclysmic binaries. In addition, Ritter (1976) has suggested that it is possible for widely separated binaries (necessary to develop a massive carbon-oxygen core) to lose enough angular momentum during their evolution to become close binaries.

The second problem has been to devise a scheme that will mix carbon nuclei from the core into the envelope. We have previously discussed possible mechanisms for this mixing (SST) but all have seemed rather unlikely on various grounds. A new mechanism has recently been, suggested based on the work of Lamb and Van Horn (1975) who found that as a pure carbon white dwarf evolved to lower effective temperatures it would develop a convective region near the surface. Colvin, et.al. (1977) then used one of Lamb and Van Horn's models to study·the effect of hydrogen accreting onto the surface. They found that protons could penetrate to the bottom of the convective zone and produce a region with both carbon and hydrogen nuclei present. However, the change in opacity as the protons were added to the white dwarf caused the bottom of the convective region to gradually move outward in mass fraction. The final result was a carbon core, a region where the percent of hydrogen varied from zero to its solar value, and the surface layers which had a solar composition.

However, because the convective region disappears during the accretion phase, the carbon nuclei that end up in the envelope are only those that were part of the initial convective zone. In the one model studied by Colvin et.al.,the convective zone was not large enough to produce a significant enhancement of the envelope. Nevertheless, white dwarfs with lower masses develope larger convective zones (Fontaine and Van Horn, 1976) and accretion onto these stars should result in a larger number of carbon nuclei in the envelope.

We have recently found evidence for another enhancement mechanism that could add carbon nuclei to the envelope during the actual outburst (Starrfield, et.al., 1977). We found that in models with low envelope mass ($M \sim 10^{-4} M_\odot$), the expansion of the envelope during the rise to peak temperature caused the outer edge of the core to expand and cool. The outermost zones of the core expanded more rapidly and quickly became convective. Although, we cannot expect mixing up through a composition discontinuity, it certainly seems possible that convective overshooting could drive some carbon nuclei into the shell source at a time when the added fuel would have its greatest effect. We have not tried to model this behavior, as yet, but will do so in the near future.

III.   OBSERVATIONS OF ENHANCED CNO NUCLEI

     This seems an appropriate place to remind the reader that en-
hanced CNO nuclei have been observed in novae ejecta for some time;
although, only qualitative estimates have been made of the degree of
enhancement. It was noticed by McLaughlin (1936) more than 40 years
ago that at maximum the principal spectrum of a fast nova showed
strong lines of CI and OI, but was otherwise almost normal. Then
Payne-Gaposchkin (1958) made the following comments in her Handbuch der
Astrophysik article on Novae (page 762):

     "The composition of novae is an important and difficult prob-
     lem. The absorption spectra of many, though not all, show
     anomalies when compared with those of other stars of similar
     dimensions, such as the yellow supergiants. Absorption lines
     of the atoms of carbon, nitrogen and oxygen are sometimes
     abnormally prominent...

     It would be premature to deduce differences of composition
     from the intensities of these forbidden emission lines,
     when the details of the excitation process are not fully
     understood. The absorption spectra are more promising,
     though here again we cannot rule out peculiar excitation
     mechanisms. The author has made a quantitative study of
     DQ Herculis, and reaches the conclusion that (if conven-
     tional curve-of-growth procedures are valid) the lighter
     elements are present in high abundance. Relative to the
     solar abundances, the numbers of atoms that contribute to
     the absorption spectra of DQ Herculis appear to fall off
     rather steadily with increasing atomic number. This result
     is no more than suggestive until analyses of other novae
     have been made. The spectrum of RR Pictoris does not show
     such high relative abundances of carbon, nitrogen and oxy-
     gen as that of DQ Herculis, but the latter star does indeed
     appear to be somewhat metal-poor."

More recently Nova Cygni showed strong oxygen and carbon lines in its
pre-maximum spectrum. In SST we discuss some of the observations of
abundances in the nebular shell which also suggest that C, N, and O
are overabundant in some novae ejecta.

     By far the most convincing evidence that a thermonuclear runaway
is the cause of the classical nova outburst was presented by Sneden
and Lambert (1975). Unfortunately, as has already been pointed out
(Sparks, et.al. 1976c),the conclusions of their paper disagree with
their text. When DQ Her was first observed it showed strong, wide,
emission lines of H, HeI, CII, NII and other elements; in addition,
very strong CN absorption appeared near maximum and lasted for a few
days (McLaughlin, 1960). Sneden and Lambert (1975) analyzed the CN
bands and found that they were unable to fit the profiles without add-
ing a large amount of $^{13}C^{14}N$ and possibly some $^{12}C^{15}N$. They state
that ^{13}C is definitely present and enhanced and ^{15}N possibly present

and overabundant. However, because their upper limits disagreed with
the numerical predictions of Starrfield, et.al. (1974), they stated in
their conclusions that there was a significant disagreement between our
results and the observations. This may be true, but the disagreement
is certainly caused by the fact that the gross features and final isotopic predictions of the outburst depend on the mass of the white
dwarf, the envelope mass, and the degree of enhancement and we have not
yet attempted to specifically model the behavior of DQ Her. The fact
is that not only are the $^{12}C/^{13}C$ and $^{14}N/^{15}N$ ratios they publish (as
upper limits) far above solar but, also, they are very different from
equilibrium CN burning predictions. This strongly supports our model
for the outburst.

IV.   THE EVOLUTIONARY STUDIES

We have previously presented the results for models with enhanced
abundances (SST). However, the abundance predictions were not reviewed and those results will be presented here. In addition to the
models with enhanced abundances, we have recently evolved models with
no CNO enhancement in a successful attempt to model the behavior of the
slow nova outburst (Sparks, et.al. 1976a, 1976b). We find that these
models reproduce the gross features of the thermonuclear runaways in
the enhanced models but no mass ejection occurs as a direct result of
the initial burst of energy generation. Nevertheless, mass ejection
does occur at a later time during the outburst because the luminosity
reaches the Eddington limit and radiation pressure drives the surface
layers off the white dwarf at a time when the layers extend to very
large radii. Such behavior was recently proposed by Bath and Shaviv
(1976) and an attempt to model it was also done by Prialnik, et.al.
(1976, preprint). The isotopes ejected are listed in Table 1 under
Model 1. Recently, some of us have studied these models for pulsa-
tional instability and find that they become unstable at a late stage
in the outburst. It is possible that pulsations in a star that fills
its Roche Lobe could also drive mass loss. These results are very
important since they show that a nova like outburst, including mass
ejection, occurs at all degrees of CNO enhancement and certainly
strengthens our theory that the classical nova outburst is caused by a
thermonuclear runaway in the envelope of a white dwarf.

The next three models that I would like to discuss all are $1.00M_\odot$
models with carbon enhanced envelopes. They have identical luminosities,
$L=2\times10^{-2}L_\odot$, but different envelope masses: $10^{-4}M_\odot$ (Model 2), $4\times10^{-4}M_\odot$
(Model 3), and $1.26\times10^{-3}M_\odot$ (Model 4). Since each model extends to a
different mass fraction in the white dwarf, the temperature (T_{tr}) and
density (ρ_{TR}) at the composition discontinuity vary by large factors
from model to model. Model 4 is the most degenerate at the composition
discontinuity, Model 2 is the least. The difference in initial temperature results in very different growth times for the thermonuclear runaway (τ_{30} is the time for the model to evolve to a shell source temperature of 30 million degrees). The temperature variation of the shell
source in Model 2 is shown in Figure 1, that of Model 4 is shown in

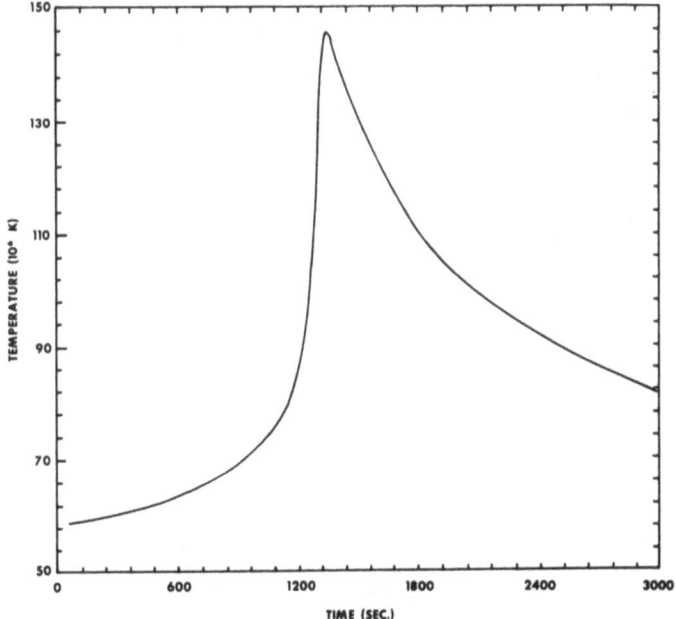

Figure 1: The temperature at the composition interface as a function of time for Model 2.

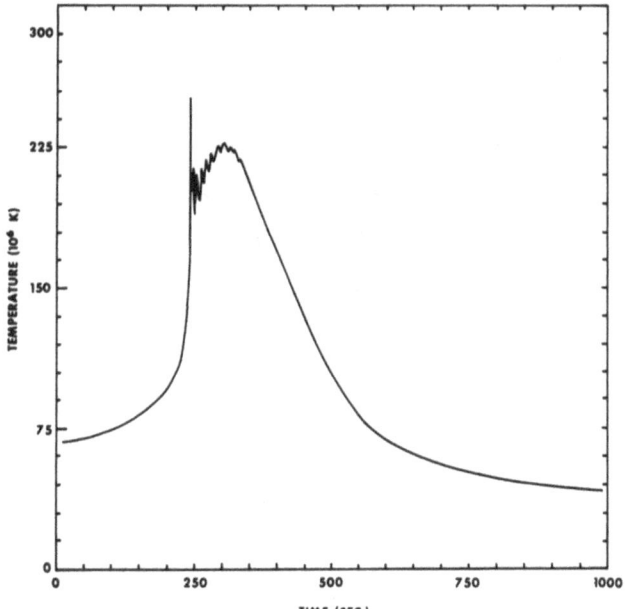

Figure 2: The temperature at the composition interface as a function of time for Model 4. The strong initial pulse produces a shock wave and the oscillations are caused by actual mass motions in the envelope. Note the difference in time scale and peak temperature between Model 2 and Model 4.

TABLE 1

RESULTS OF THE EVOLUTION

MODEL	1	2	3	4	5
a) Initial Conditions					
$L/L_\odot$	2.9-3	1.8-2	1.8-2	1.8-2	1.8-2
$\text{Log } T_e$	4.16	4.30	4.30	4.30	4.30
$R(km)$	6054	7755	7589	7976	7978
$M_e(M_\odot)$	1.1-4	1.1-4	4.1-4	1.3-3	1.3-3
$\text{Log } T_{TR}$	6.939	7.086	7.178	7.232	7.238
$\text{Log } \rho_{TR}$	3.755	3.386	3.801	4.148	4.147
$\tau_{30}(yrs)$	5.4+5	1.0+3	16.	1.19	1.40
$^{12}C(ADDED)(gm)$	0.0	1.0+29	1.0+29	1.5+29	1.0+29
$^{16}O(ADDED)(gm)$	0.0	0.0	0.0	0.0	1.0+29
b) Final Conditions or Peak Conditions					
$T_{peak}(10^6 K)$	153	146	199	321	252
$\varepsilon_{nuc}(\text{erg gm}^{-1}\text{s}^{-1})$	6.6+13	4.2+15	3.8+17	1.0+18	1.8+17
MASS EJ. (gm)	N.A.	7.0+28	2.3+28	2.3+28	2.2+29
$V_{min}(km/s)$	N.A.	32	135	335	354
$V_{max}(km/s)$	N.A.	3217	2850	5738	2888
$M_{Bol}(MAX)$	-6.6	-11.3	-10.1	-8.9	-9.1
$M_{Vis}(MAX)$	0.2	-7.5	-6.2	-6.8	-7.5
c) Ejected Isotopes – Percent by Mass					
^{12}C	3.9-4	1.3-1	4.1-3	1.2-3	7.4-3
^{13}C	1.2-4	2.3-1	2.1-1	1.6-2	5.7-3
^{14}N	1.0-2	1.2-1	8.4-2	3.6-2	3.1-2
^{15}N	3.1-7	9.8-4	3.6-2	1.8-2	4.5-2
^{16}O	1.1-4	7.1-3	6.4-3	6.7-3	8.4-3
^{17}O	2.4-8	6.0-5	7.0-4	5.1-4	1.6-2
^{7}Li	N.A.	1.1-6	6.7-6	2.4-6	N.A.

N.A. = NOT AVAILABLE

Figure 2.   Model 4 reaches much higher temperatures (T peak) than either models 2 or 3 because it is initially more degenerate (at the composition discontinuity), and because it has 50% more carbon nuclei than
the other two models.   The sharp peak in the temperature (Model 4) is
the result of the formation of a shock which moves through the envelope
in a few seconds and ejects a small amount of material (Sparks 1969).
Neither of the other models produces a shock.   At maximum temperature
the abundances by mass of the β^{+}-unstable nuclei exceed 10^{-2}.   A graph
showing their behavior as a function of time appears in SST.

These models eject around $10^{-5} M_{\odot}$ moving with speeds from a few
hundred to a few thousand km/sec.   The light curves for models 2, 3,
and 4 are given as Figures 3, 4, and 5.   It is quite clear that the
envelope mass has a strong effect on the light curve.   Notice the very
rounded maximum for Model 2.   It is very reminiscent of Nova Cygni
1975.   There is also a large variation in the peak visual magnitude.
In fact, the model with the smallest envelope mass, Model 2, reaches
the largest magnitude.

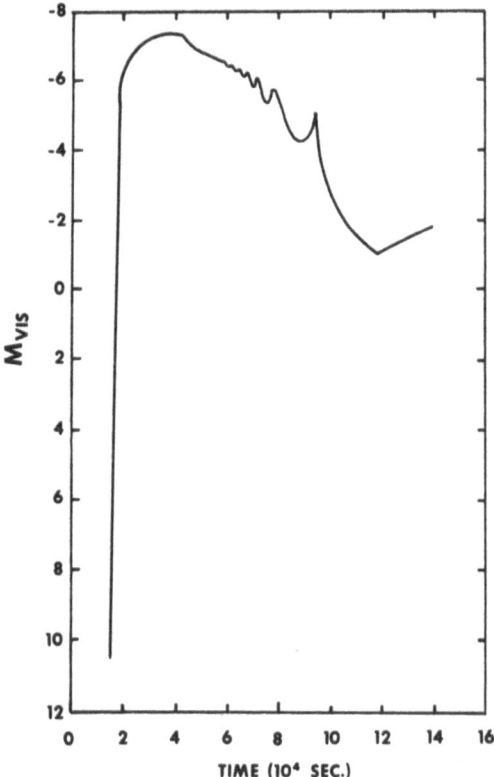

Figure 3:   The visual light curve for Model 2.   Note the rounded maximum and time scale for this model compared to Models 3 and 4.

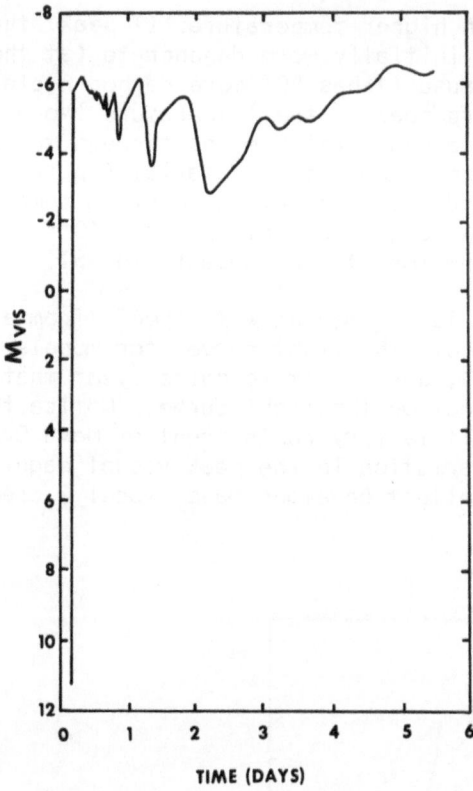

<u>Figure 4</u>:   The visual light curve for Model 3.

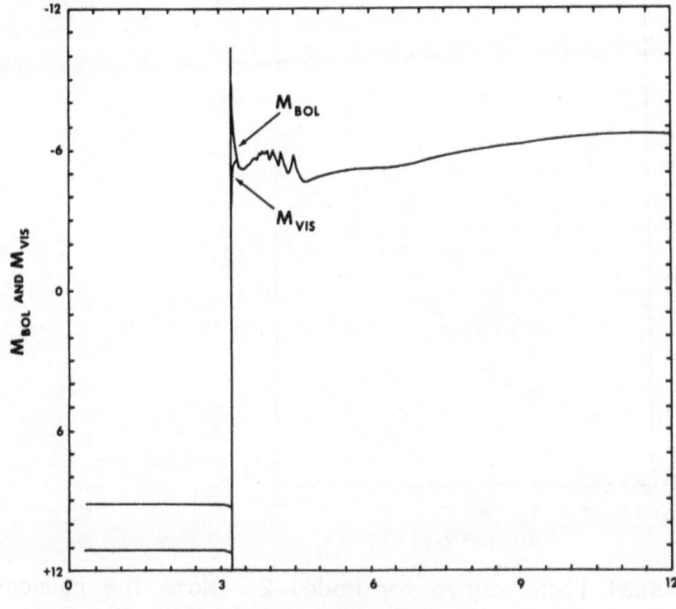

<u>Figure 5</u>:   The bolometric and visual light curve for Model 4.

Table 1 also shows that the abundances in the ejected material depend on the envelope mass although other factors, i.e. white dwarf mass, white dwarf luminosity, and CNO enhancement must also affect the abundances. In order to show this we also include the results for one previously published model (SST). The $^{12}C/^{13}C$ ratio is a very strong function of envelope mass, varying from .6 in Model 2 to 0.075 in Model 4. It can also be seen that it depends strongly on carbon and oxygen enhancement. Model 1 which was not enhanced has a $^{12}C/^{13}C$ ratio of 3 while Model 2 shows a $^{12}C/^{13}C$ ratio of 0.6. The $^{14}N/^{15}N$ ratio also varies with the above considerations from a value of 10^5 in Model 1 to a value of 2.0 in Model 4 and 0.7 in Model 5. Note that the only difference between Models 4 and 5 is the initial values of ^{12}C and ^{16}O in the deepest hydrogen rich shell. It is clear that these variations alone have a large effect on the final abundances with $^{12}C/^{13}C$ varying from less than 1.0 to greater than 1 and the reverse being true for the $^{14}N/^{15}N$ ratio. In addition, the total N/C ratio may give us some information on the envelope mass since N/C is greater than 1 for those enhanced models with $M_e > 4 \times 10^{-4} M_{\odot}$ and less than 1 in Model 2 (and other unpublished low envelope models).

The last line presents the percent by mass of ^7Li ejected by the nova. This is a new result and is based on the previously mentioned expansion of our network for the proton-proton chain. Arnould and Nørgaard (1975) showed that the temperature and density conditions that we found in the nova outburst were sufficient to convert some fraction of the initial ^3He into ^7Li. The ^7Li abundances in Table 1 show that their conjecture was correct! The novae convert anywhere from 10 to 50 percent of the initial ^3He to ^7Li. In fact, the abundance of ^7Li is so large that the novae may be responsible for a significant amount of galactic ^7Li.

V. SUMMARY

We have shown in our published work that a thermonuclear runaway in the hydrogen envelope of a carbon-oxygen white dwarf can reproduce the gross features of the classical nova outburst. It is also apparent that the behavior of the outburst depends upon (at least) the hydrogen envelope mass and the degree of enhancement. It has also been shown that all degrees of isotopic enhancement result in an outburst that may be compared with observed events.

Our predicted isotopic ratios for the ejected material vary considerably as a function of the parameters that we have discussed. Although, we have not yet produced a model which reproduces the isotopic upper limits of Sneden and Lambert (1975), the variations in our published isotopic ratios include their limits; and it should be possible to provide such a model, especially if we include mixing with unprocessed material in our calculations (circumstellar or circumbinary).

It now appears that we have isolated possible enhancement mechanisms. This is doubly important since the observational evidence

strongly suggests that the CNO nuclei are enhanced in novae ejecta.
The two mechanisms that we have described both involve mixing in the
envelope. The first one is based on the work of Lamb and Van Horn
(1974) who found that a carbon white dwarf developed a convective zone
near its surface. A study of hydrogen accreting onto this white dwarf
then showed that a region of carbon plus hydrogen nuclei would result
(Colvin, et.al. 1977). The second mechanism is based on our recent
hydrodynamic studies of low envelope mass outbursts. We find that the
outer edge of the carbon core becomes convective during the outburst
and any convective overshoot will inject carbon nuclei into the shell
source just after it has reached peak temperature.

    We would like to acknowledge valuable discussions with J. Galla-
gher and J. Audouze on CNO nucleosynthesis. We are grateful to the
National Science Foundation for support under Grants AST73-05271-A01
(SGS) and AST73-05117-A02 (JWT). S. Starrfield would like to thank
Dr. Peter Carruthers for the hospitality of the Los Alamos Scientific
Laboratory and a generous allotment of computer time.

REFERENCES

Arnould, M. and Nørgaard, H. 1975, Astr. and Ap., 42, 55.
Bath, G.T., and Shaviv, G. 1976, M.N.R.A.S., 175, 305.
Colvin, J. D., Van Horn, H. M., Starrfield, S. G., and Truran, J. W.,
    1977, Ap. J., in the press.
Fontaine, G., and Van Horn, H. M., 1976, Ap. J. Suppl., 31, 467.
Giannone, P., and Weigert, A. 1967, Zs. f. Ap., 67, 41.
Kutter, G. S., and Sparks, W. M. 1972, Ap. J., 175, 407.
Lamb, D. Q., and Van Horn, H. M. 1975, Ap. J., 200, 306.
McLaughlin, D. B. 1936, Ap. J., 84, 104.
McGlaughlin, D. B. 1960, Stars and Stellar Systems VI. Stellar
    Atmospheres, p 585, ed. J. L. Greenstein, University of Chicago
    Press.
Payne-Gaposchkin, C. H. 1958, Handbuch d. Physik LI, p 752, ed. S.
    Flügge, Springer-Verlag, Berlin.
Prialnik, D., Shara, M. M., and Shaviv, G. 1976, preprint.
Ritter, H. 1976, M.N.R.A.S., 170, 533.
Sparks, W. M., Starrfield, S. G., and Truran, J. W. 1976a, Bul. A.A.S.,
    8, 321.
Sparks, W. M., Starrfield, S. G., and Truran, J. W. 1976b, Proceedings
    of the Paris Nova Meeting, to be published.
Sparks, W. M., Starrfield, S. G., and Truran, J. W. 1976c, Ap. J., 208,
    819.
Starrfield, S. G., Sparks, W. M., and Truran, J. W. 1974, Ap. J. Suppl.,
    28, 247.
Starrfield, S. G., Sparks, W. M., and Truran, J. W. 1976, I.A.U. Sympo-
    sium No. 73, Structure and Evolution of Close Binaries, ed. P.
    Eggleton, S. Mitton, J. Whelan, Reidel, Dordrecht (SST).
Starrfield, S. G., Sparks, W. M., and Truran, J. W. 1977, in prepara-
    tion.
Starrfield, S., Truran, J. W., Sparks, W. M., and Kutter, G. S. 1972,

Ap. J., 176, 169.
Taam, R., and Faulkner, J. 1975, Ap. J., 198, 435.
Warner, B. 1976, I.A.U. Symposium No. 73, Structure and Evolution of
    Close Binaries, ed. P. Eggleton, S. Mitton, J. Whelan, Reidel,
    Dordrecht.
Webbink, R. 1976, preprint.

THE CNO COMPOSITION IN NOVAE ENVELOPE

D. Prialnik, M.M. Shara and G. Shaviv
Department of Physics and Astronomy, Tel-Aviv University,
Ramat-Aviv, Israël.

The fashionable nova model today is one in which hydrogen rich matter (i.e. normal cosmic composition) accreted by a White Dwarf gives rise to a thermonuclear runaway and ejection of a sizeable fraction of the envelope. Recently extensive numerical calculations of this model were carried out by Starrfield, Sparks and Truran (1974) (SST). In these calculations the assumed initial model is a degenerate carbon core of $1M_\odot$. with an envelope of $(1.25 - 1.7) \times 10^{-3} M_\odot$. The most important result discovered by SST in this context of CNO composition in astrophysics is the following : In order to achieve a mass loss of the order observed ($\sim 10^{-4} M_\odot$) it is necessary to assume that somehow the envelope is enriched by CNO. The total mass fraction of CNO is the models run by SST is ≥ 0.3. If this enrichment is not assumed the outcome of the runaway is a semi red giant configuration with radius of $14R_\odot$, luminosity of $\sim 1.5 \times 10^4 L_\odot$ and nuclear burning timescales of $10^3 - 10^4$ years.

The assumption of high CNO abundance has important implications on the composition of the ejected material. The ejected material should have $[C/C_\odot] \gg 1$, $[N/N_\odot] \gg 1$ and $[0/0] \gg 1$. Observations by Pottanch (1959) and the analysis by Collin-Souffrin (1976) seem to confirm this prediction only partly. (The observation of carbon is extremely difficult and unreliable at the present).

We report here on different nova models calculated by us. We assume normal CNO abundance and an envelope mass of $10^{-4}M_\odot$. The envelope is in local thermodynamic equilibrium and the

Jean Audouze (ed.), CNO Isotopes in Astrophysics, 63-67. All Rights Reserved.
Copyright © 1977 by D. Reidel Publishing Company, Dordrecht-Holland.

whole star in thermal equilibrium. In view of the theoretical
importance of possible mixing between the interior and the
envelope we have included the whole star in our calculations.
The model takes about 10^4 yr to flash and reaches burning
temperatures of 1.5×10^8 K with peak energy generation of
5.4×10^{13} erg gm^{-1} sec^{-1}. The time spent by burning shell
at $T > 10^8$ K is 3×10^5 sec. Consequently, the CNO cycles had
enough time to reach equilibrium. At the peak of the flash
the nuclear luminosity rises to $5.5 \times 10^8 L_\odot$. The outer layers
start to move outward and the bolometric luminosity start to
rise quite quickly several days after the maximum in the
nuclear luminosity.

Mass loss starts as soon as $L_{Bol} \sim \cdot 7$ L Eddington. A
steady slow mass loss appears first and later a steady fast
mass loss develops. The bolometric luminosity decreases on
a time scale of months, i.e. $L_{Bol} \sim 10^{-3} L_\odot$ 2 years past the
flash. During the whole phase of mass loss the bolometric
luminosity remains constant in agreement with the observation
of Gallagher and Code (1974) and it starts to decline only when
95% of the envelope is ejected and mass loss stops. Our model
predicts therefore the most important features of the nova
phenomenon, namely, steady state outflow, $L_{Bol} \approx$ const for
several months, as soon as $\sim 10^{-4}$ $M_\odot$ are ejected the bolometric
luminosity declines, and the time dependence of the visual
luminosity agrees well with the classical slow nova behaviour.
Several other features like velocity of expansion, time between
peak nuclear luminosity and peak visual luminosity etc agree
as well with the observations.

The results for the isotopic compositions to be found in
the ejected material are given in table 1. It should be noted
that the ejecta composition in terms of total amounts of carbon
say, compared to the solar composition as well as the isotopic
ratio is (a) very sensitive to the details of the model and (b)
is time dependent. The exact time history of convection and
its extent for example will affect the isotopic ratio (how deep
the convection penetrates into the burning shell and how far
it extends into the outer layers). The variability of the
composition with mass ejected is demonstrated in Table 2. The
effect of the total mass ejected on the composition is clearly
seen. Hence, the reliability of the predicted composition
depends on the ability of the model to describe properly other
observed features.

Of special interest are the isotopic ratios of C^{12}/C^{13},
N^{14}/N^{15} observed by Sneden and Lambert (1975). While the
observation of C/H or N/H is very complicated and to some
extent model dependent, the isotopic ratio is expected to be
less model dependent and more reliable (if the two isotopes

Table 1

Comparison between the observed and predicted CNO abundances in Nova ejectas.

Author	Work	$C/C_\odot$	$N/N_\odot$	$O/O_\odot$	C^{12}/C^{13}	N^{14}/N^{15}	O^{16}/O^{17}
Pottasch (1959)	Observations of 5 Novae	$1^{+}_{-0.5}$	45^{+}_{-15}	5^{+15}_{-4}	-	-	-
SST (1974)	Nova model (prediction)	4	60	3	$\frac{1}{2} - \frac{1}{7}$	$\frac{2}{3}$	4
Collin-Souffrin	Nova Herculis (1963) (observations)	-	50^{+150}_{-30}	10^{+30}_{-6}	-	-	-
Sneden & Lambert (1975)	DQ Her (1934) (observations)	-	-	-	$\gtrsim 1.5$	$\gtrsim 2$	-
Present work	Nova model (prediction)	0.3	18	1.8	2.8	3.8×10^4	3.3×10^3

Table 2

The composition of ejected material as a function of ejected matter (in per cent of the total mass of the original envelope).

%	H^1	He^4	He^3 $\times 10^6$	C^{12} $\times 10^4$	C^{13} $\times 10^4$	N^{14} $\times 10^2$	N^{15} $\times 10^7$	O^{16} $\times 10^2$	O^{17} $\times 10^6$
5	0.676	0.294	0.999	5.88	1.95	1.66	4.75	1.31	2.34
5	"	"	1.05	6.12	2.04	1.65	4.68	1.31	2.74
52.25	"	"	1.15	6.59	2.24	1.65	4.39	1.31	3.59
4.25	"	"	1.35	7.15	2.61	1.63	4.30	1.32	4.52
10.0	"	0.293	1.41	7.39	2.79	1.62	4.24	1.32	4.87
12.5	"	"	1.78	8.00	3.06	1.62	4.08	1.32	4.82
2.5	"	"	2.34	8.91	3.46	1.61	3.83	1.33	4.75
3.5	"	"	3.95	1.16	4.48	1.59	2.16	1.34	4.21
Average for 95%	0.676	0.294	1.39	7.07	2.51	1.64	4.30	1.31	3.87

are not too far apart on the curve of growth).

We conclude therefore that our model with initial "solar composition" predicts the observed hydrodynamic phenomena as well as the composition ratios found in the expanding nebula.

REFERENCES

Starrfield S., Sparks W.M., and Truran J.W. 1974, Ap. J.
             Suppl. Ser. No 261, 28, 247
Pottasch S. 1959, Ann. d'Ap. 22, 412
Collin-Souffrin S. 1976 (preprint)
Gallagher J.S. and Code A.D. 1974, Ap. J. 189, 303.

PART IV

CNO ISOTOPES IN THE INTERSTELLAR MEDIUM

CNO ISOTOPE ABUNDANCES IN INTERSTELLAR CLOUDS

Peter G. Wannier
California Institute of Technology

1. INTRODUCTION

Isotope abundance determinations have been made to date in at least twenty-three dense clouds scattered throughout the galactic disc. Most of these determinations have been made using two or more of the sixteen observed isotopic species of CO, H_2CO, HCN and $CS^{\dagger}$. As will be shown, only a few of the molecular experiments can be interpreted independently. The two most notable of these independent experiments are the comparisons of the 6.2 cm absorption lines of H_2CO and $H_2^{13}CO$ and 2.6 mm emission lines of $C^{18}O$ and $C^{17}O$. Both of these experiments will be discussed at length in subsequent articles in these proceedings. The difficulty of obtaining single isotope abundance ratios is an experimental one. Most terrestial ratios are very large, ranging up to 2675 for the ratio ^{16}O / ^{17}O. These large ratios create an experimental bind. Spectral lines of the more abundant isotopic species must be weak enough to ensure that the line suffers no serious optical depth effects. This requirement often makes the line of the less abundant species impossible to detect. Hence many experiments yield double isotope abundance ratios by comparing the line strengths if two molecules having isotope substitutions of two different elements (e.g., $|C^{18}O|/|^{13}CO|$ which yields $(^{12}C/^{13}C)/(^{16}O/^{18}O)$).

A fruitful approach is to concentrate on those dense clouds where several isotope abundance experiments have been done. This approach is also the best one to control for any errors, either experimental or resulting from optical depth effects and chemical fractionation. Table 1 presents the experimental results from eleven dense clouds, all those to date where two or more experiments have yielded significant results. The numbers presented are the molecule abundance ratios written as a percent of the corresponding terrestrial ratio. The terrestrial ratios which have been used are presented in the last row

---

†Hence forth I shall adopt the usual convention that any molecule, unless otherwise specified, is comprised of its most abundant nuclear species. Thus $^{13}C^{16}O$ is written ^{13}CO and $^{1}H^{12}C^{14}N$ is HCN.

Table 1

Source	R_c	$\dfrac{C^{18}O}{^{13}CO}$	$\dfrac{C^{34}S}{^{13}CS}$	$\dfrac{H_2CO}{H_2{}^{13}CO}$	$\dfrac{C^{17}O}{C^{18}O}$	$\dfrac{C^{33}S}{C^{34}S}$	$\dfrac{HC^{15}N}{H^{13}CN}$	$\dfrac{DCN}{H^{13}CN}$
SGR A	0.1			24(12) [9]			2(5) [7]	
SGR B	0.2	22(10) [1]	43 (6) [6]	20(10) [9]	168(19) [2]	105(25) [6]	6(5) [7]	2(1) [8]
W51	8.0	<84 [1]	68(17) [6]	42 (2) [5]			63(5) [7]	13(3) [8]
M17	8.0	61 (3) [1]	42 (5) [6]					13(4) [8]
NGC6334	9.3	<67 [1]		55 (4) [5]	162(19) [2]		39(8) [7]	10(2) [8]
DR21	9.9	28 (6) [1]			114(12) [2]			
CLOUD 4	10.0	50 (7) [4]			124(12) [2]			
ORION A	10.9	42 (2) [1]	68 (8) [6]	40(10) [1]	146(36) [3]	116(25) [6]	76(6) [7]	17(2) [8]
NGC2024	10.8	24 (6) [1]		48 (6) [5]	108 (8) [2]			
NGC2264	11.1	46(14) [1]	70(12) [6]	>36 [5]	92(17) [2]			
NGC7538	12.7	42 (3) [1]			108(12) [2]		50(9) [7]	
Terr. Ratios	10.0	.178	3.85	89.	.185	.18	.32	(1.00)

1. Wannier et al., 1976a
2. Wannier et al., 1976b
3. Encrenaz et al., 1973
4. Encrenaz, 1974
5. T. L. Wilson et al., 1976
6. R. W. Wilson et al., 1976
7. Linke et al., 1976
8. Penzias et al., 1976
9. Whiteoak & Gardner, 1972

of the table and are taken from Wedepohl (1969) except D/H for which
no meaningful (unfractionated) data exist from anywhere in the solar
system. The uncertainties presented in Table 1 have been reevaluated
by the author and include not only experimental uncertainties but also
the estimated uncertainties resulting from optical depth effects and
calibration uncertainties. All quoted errors are ± 2σ.

The possible role of chemical fractionation in causing the molecule abundance ratios to differ from the total isotope abundance ratios
will not be discussed in detail. This subject, which has recently been
studied, will be treated by a subsequent article in these proceedings.
In this paper the approach will be to observe several different molecules and look for consistent results. Thus the $^{12}C/^{13}C$ ratio will be
examined directly in each of four molecules.

All carbon monoxide data and all hydrogen cyanide data except for
DCN are from the J = 1→0 emission lines. The carbon monosulphide data,
the DCN data and the formaldehyde data in Orion A are from the J = 2→1
emission lines. All formaldehyde data except in Orion A are from the
$1_{10}→1_{11}$ transition at 6.5 cm. The number R_c in column 2 is the galactocentric distance in kpc.

2. DISCUSSION

Such a table can be viewed in several different ways to yield
useful information about nucleosynthesis and galactic chemical evolution. First, a quick glance over all the numbers in the table can
indicate a possible evolution of the interstellar medium since the
time when the protosolar nebula itself formed some five billion years
ago. Second, a glance down the columns of the table may indicate
radial abundance gradients associated with the different stellar populations at various distances from the Galactic Center. Third, and
finally, cloud-to-cloud variations at comparable galactocentric radii
might indicate recent (local) changes, possibly from some single event
or set of events such as from a supernova or mass loss from a cluster
of stars. These three viewpoints will be discussed in order.

2.1 Chemical Evolution in the Solar Neighborhood

To the end of discussing chemical evolution we consider the overall values in table 1 and especially those from clouds with R_c between
8 and 12 kpc, comparable to the Sun. This condition includes all regions except for the two Galactic Center sources. Because of the
wealth of organic molecules in dense clouds, the carbon isotopes are
the best studied. Columns 1, 2, 3, 6 and 7 of the table all involve
the $^{12}C/^{13}C$ either directly or indirectly as part of a double ratio.
Because of the special status of the $|DCN|/|H^{13}CN|$ experiment in
column 7, it will be considered separately. The single most general
result is that all of these ratios in all regions included in the
table are significantly (2σ) less than the corresponding terrestrial
abundance ratios. The simplest explanation for the set of results

is that the $^{12}C/^{13}C$ ratio has been reduced, more likely by the addition the rare ^{13}C isotope than by the more difficult net destruction of ^{12}C. This explanation is further supported by considering each of the four studies in greater detail.

Of these experiments, only the $|H_2CO|/|H_2^{13}CO|$ provides a direct measure of $^{12}C/^{13}C$. Effects of line saturation in the $|H_2CO|/|H_2^{13}CO|$ experiments would lower the observed values, an effect which is certainly important in Sgr B and Sgr A where the quoted uncertainties result from unknown optical depth. The ratio always seem to be smaller than terrestrial by factors of two or more. The other three experiments deal with double isotope ratios where saturation effects are generally less severe but where interpretation in terms of single ratios is more involved.

The $|C^{18}O|/|^{13}CO|$ experiment (column 1) simultaneously measures $^{12}C/^{13}C$ and $^{16}O/^{18}O$. Complete deconvolution of these ratios may come about by measurement of $|C^{18}O|/|^{13}C^{18}O|$, a difficult experiment which has been performed in NGC 2024 yielding a value of about half the terrestrial one (Wannier et al., 1976a). The results of the $|C^{17}O|/|C^{18}O|$ experiment (column 4) certainly do not indicate any significant underabundance of ^{18}O, strengthening the case for a low $^{12}C/^{13}C$ ratio.

The $|C^{34}S|/|^{13}CS|$ experiment likewise produces ratios significantly less than terrestrial, but simultaneously involves the $^{34}S/^{32}S$ isotope ratio. Measurements of $|C^{33}S|/|C^{34}S|$ (column 5) in two regions yield terrestrial abundances for these two rare sulphur isotopes. This result is in accordance with a model of sulphur production during explosive nucleosynthesis (Arnett and Clayton, 1970) which predicts that ^{32}S, ^{33}S and ^{34}S should all end up in their terrestrial ratios. The experimental results that the $|C^{34}S|/|^{13}CS|$ ratios are low by a factor two is in good agreement with a low $^{12}C/^{13}C$ ratio (also by a factor two) and a terrestrial value for interstellar $^{34}S/^{32}S$.

Finally, the $|HC^{15}N|/|H^{13}CN|$ experiment simultaneously measures $^{12}C/^{13}C$ and $^{15}N/^{14}N$ (column 6). Except in the Galactic Center, the double ratio is again about half of its terrestrial value. Independent evidence for a sub-terrestrial $^{12}C/^{13}C$ ratio comes from a detailed analysis of the HCN/H^{13}CN intensity ratio in positions south of Orion A. Because T_A^*(HCN) is much less than the kinetic temperature, it can be shown that $^{12}C/^{13}C$ lies between 30% and 65% of its terrestrial value (Linke et al., 1976; Wannier et al., 1974).

It is interesting to note that any effects of line saturation in the $|C^{18}O|/|^{13}CO|$ experiment and in the $|HC^{15}N|/|H^{13}CN|$ experiment would tend to <u>increase</u> the observed ratios, the opposite effect from that in the $|H_2CO|/|H_2^{13}CO|$ experiment or in the $|C^{34}S|/|^{13}CS|$ experiment. Also, possible chemical fractionation schemes which involve carbon dioxide leave the hydrogen cyanide molecule unfractionated, or at least heavily fractionated relative to carbon monoxide. These two points demonstrate very clearly the advantages of performing several different experiments in each cloud. As can be seen from the above remarks, the values in table 1 from the region of $R_c \approx 10$ kpc are in general agreement with solar values for all measured isotope ratios except for $^{12}C/^{13}C$ which appears to be about a factor two less. These other measured ratios include $^{18}O/^{16}O$, $^{17}O/^{18}O$, $^{34}S/^{32}S$, $^{33}S/^{34}S$ and $^{15}N/^{14}N$. Inasmuch as the galactic disc between 8 kpc and 12 kpc from the Galactic Center can be thought to typify the solar neighborhood by constituting a common nucleogenic pool, the $^{12}C/^{13}C$ ratio appears to have evolved away from the value of the pool since the birth of the sun.

2.2 Radial Abundance Gradients

The large number of sources between 8 and 12 kpc, which aided the study of chemical evolution in the solar neighborhood, make the study of radial abundance gradients difficult. The selection is a natural result of observing from telescopes in the northern hemisphere where clouds with $R_c \lesssim 6$ kpc all have low declinations and conflict in observing time with the Galactic Center sources themselves. The data in table 1 thus fall into two parts, those from the Galactic Center (Sgr A and Sgr B) and those with approximately solar galactocentric radii. There are at least two values in table 1 which indicate strong differences between the "Center" sources and the "outer" sources, both using isotopically substituted hydrogen cyanide.

In order to interpret the $|HC^{15}N|/|H^{13}CN|$ results we first examine the evidence for a $^{12}C/^{13}C$ abundance gradient. The $|H_2CO|/|H_2^{13}CO|$ survey yields a line intensity ratio of about 10 in the Center regions. A complete analysis of H_2CO opacity has been made by Whiteoak and Gardner (1972). The opacity results in a range of $^{12}C/^{13}C$ in the Center from about the same as in the outer sources to about a factor three less. Another measure of $^{12}C/^{13}C$ is from the $|C^{18}O|/|^{13}CO|$ experiment. The abundance ratio in Sgr B is again smaller than in the outer sources but with a range similar to that given above. The $|C^{34}S|/|^{13}CS|$ experiment yields a ratio not significantly different from in the outer regions, but again slightly smaller. If we accept the conjecture that $^{34}S/^{32}S$ should be terrestrial, then we get a $^{12}C/^{13}C$ ratio 1 to 2 times less than in the outer regions.

Thus, the $^{12}C/^{13}C$ ratio appears to be smaller in the Center, perhaps by a factor of two and not exceeding a factor of three. This information is useful to interpret the very striking results of the $|HC^{15}N|/|H^{13}CN|$ survey in which the Center sources have values at least five times smaller than the outer sources. The upper limit in Sgr B results not from uncertain detection of the weak $HC^{15}N$ line but from the fact that the line shapes are very different for $HC^{15}N$ and $H^{13}CN$. This indicates either that: (1) the isotope ratio varies with radial velocity (and presumably, therefore, with distance) so that some positions have yet lower abundance ratios than indicated, or (2) the stronger $H^{13}CN$ line is heavily saturated so that the real abundance ratio will be lower than indicated when properly corrected.

In Sgr A, the $HC^{15}N$ line is still undetected to a very significant limit. The small double ratio can result from a low $^{12}C/^{13}C$ ratio and/or a high $^{14}N/^{15}N$ ratio. There is some evidence presented above that $^{12}C/^{13}C$ may be lower in the Center but not by the large factor necessary to produce the $|HC^{15}N|/|H^{13}CN|$ results.

The other experiment which indicates a possible abundance gradient is the $|DCN|/|H^{13}CN|$ experiment (column 7). Although chemical fractionation is virtually certain to play an important role in determining the real D/H ratio, the uniformity of the values in the outer regions and the dramatically lower values in the Center indicate a low central value of D/H, possibly combined by the somewhat uncertain reduction in $^{12}C/^{13}C$.

2.3 Cloud to Cloud Variations

Differences in isotopic composition between clouds having approximately the same value of R_c can indicate nucleosynthetic events in the recent past, possibly within the lifetime of a given cloud. If we consider as a group all of the sources outside of the Galactic Center, then there are several significant differences as seen by looking down the columns of table 1. For example, $|HC^{15}N|/|H^{13}CN|$ in Orion A yields 76% ± 6 (2σ) whereas it yields 50% ± 9 in NGC7538. The $|C^{34}S|/|^{13}CS|$ experiment yields 68% ± 17 in W51 and 42% ± 5 in M17. Several other examples can be found. The question is: do these differences reflect real isotope abundance differences or is there some additional source of error in the experiments not reflected by the formal uncertainties? It is frustrating that table 1 is not complete, so that different sources have been studied using different combinations of molecules. In the few cases where there is experimental overlap between two sources, there is no obvious detailed correlation of the results of different experiments. Thus, whereas $|HC^{15}N|/|H^{13}CN|$ is 76% ± 6 in Orion A and is 50% ± 9 in NGC7538, the $|C^{18}O|/|^{13}CO|$ results for the two sources agree exactly. The comparison of W51 and M17 cannot be extended because the two sources have no other isotope ratio in common except for the DCN results. The only conclusion that can be reached at present is that a few of the experiments give ratios which yield source-to-source variations far in excess of experimental error, indicating that recent events have had some effects. The exact nature of these variations awaits a more complete set of results.

3. CONCLUSIONS

As can be seen from the above discussion, meaningful measurements of interstellar isotope abundances can only result from an extensive program involving many molecules in many clouds. Such an approach can detect chemical evolution in the solar neighborhood, abundance gradients on a galactic scale and recent nucleosynthetic events. Comparing the results of several different experiments also serves to eliminate unsuspected errors in any one experiment, possibly from chemical fractionation, line saturation or calibration difficulties. To date, the data indicate that outside of the Galactic Center, there is a general reduction in the $^{12}C/^{13}C$ ratio by about a factor two. There is no evidence that any other of the measured isotope ratios differs greatly from the terrestrial value. There is evidence for radial abundance gradients in both $^{14}N/^{15}N$ and $^{12}C/^{13}C$, the former being high in the Galactic Center and the latter being low. Experimental results do not yet seem complete enough to make definite conclusions about cloud-to-cloud variability, but there is some evidence that abundances are not always the same among clouds having comparable R_c. More definite conclusions will surely appear as experimental results fill in the table presented.

ACKNOWLEDGEMENTS

I am indebted to Drs. Arno A. Penzias and Robert W. Wilson who helped to organize the material for this article. Partial support was provided by National Science Foundation Grant MPS 73-04908 A03.

REFERENCES

Arnett, W. D. and Clayton, D. D.: 1970, Nature 227, 780.
Encrenaz, P. J., Wannier, P. G., Jefferts, K. B., Penzias, A. A. and Wilson, R. W.: 1973, Astrophys J. 186, L77.
Encrenaz, P. J.: 1974, Astrophys J. 189, L135.
Linke, R. A., Goldsmith, P. F., Wannier, P. G., Wilson, R. W. and Penzias, A. A.: 1976, Astrophys.J. (to be published).
Penzias, A. A., Wannier, P. G., Wilson, R. W. and Linke, R. A.: 1976, Astrophys.J. (to be published).
Wannier, P. G., Encrenaz, P. J., Wilson, R. W. and Penzias, A. A.: 1974, Astrophys. J. 190, L77.
Wannier, P. G., Penzias, A. A., Linke, R. A. and Wilson, R. W.: 1976A, Astrophys. J. 204, 26.
Wannier, P. G., Lucas, R., Linke, R. A., Encrenaz, P. J., Penzias, A. A. and Wilson, R. W.: 1976b, Astrophys. J. 205, L169.
Wedepohl, K. H.: 1969, ed., Handbook of Geochemistry (Berlin: Springer-Verlag).
Whiteoak, J. G. and Gardner, F. F.: 1972, Astrophys. Letters 6, 231.
Wilson, R. W., Penzias, A. A., Wannier, P. G. and Linke, R. A.: 1976, Astrophys. J. 204, L135.
Wilson, T. L., Bieging, J. and Downes, D.: 1976, Astron. and Astrophys. (to be published).

^{17}O AND ^{18}O ABUNDANCES IN INTERSTELLAR CLOUDS

Pierre J. Encrenaz
Observatoire de Meudon, 92190 MEUDON, France ; and
Laboratoire de Physique, Ecole Normale Supérieure.

1. INTRODUCTION

The determination of isotopic abundance ratios in interstella molecules is a fast-growing subject (see the review by Wannier in this book). Because CO is so widespread in our Galaxy, its isotopic substitutions are of particular importance. We concentrate here on the two isotopic molecules $^{12}C^{17}O$ and $^{12}C^{18}O$. They give only faint J = 1 - 0 lines, at 112.36 GHz and 109.78 GHz respectively, but these lines are unsaturated even in Sgr B2 (as can be checked by comparing them to the lines of the more abundant species $^{12}C^{16}O$ and $^{13}C^{16}O$ in the same molecular clouds), and presumably their excitation conditions are similar ; thus the ratio of their intensities directly provides an abundance ratio. Whether chemical fractionation of the oxygen isotopes occurs, as it might be the case for the carbon isotopes (Watson, this book), is unknown ; thus we do not know to which extent the $^{17}O/^{18}O$ ratio is equal to the $C^{17}O/C^{18}O$ ratio.

2. OBSERVATIONS AND DATA REDUCTION

$C^{17}O$ was discovered by Encrenaz et al (1973) and since that time a more extensive survey of the $C^{17}O$ and $C^{18}O$ lines in nine dense molecular clouds has been conducted by Wannier et al (1976). This survey is the basis of the present paper. The observations, which are described in the paper by Wannier et al (1976), used the

Jean Audouze (ed.), CNO Isotopes in Astrophysics, 79-83. All Rights Reserved.
Copyright © 1977 by D. Reidel Publishing Company, Dordrecht-Holland.

11-m millimeter antenna of the NRAO at Kitt Peak (Arizona) and the
5-m antenna of the Millimeter-Wave Observatory at Fort Davis (Texas).
A special difficulty with the 11-m run came from the coincidence of
the image side-band for the $C^{18}O$ observations with a strong telluric
O_2 line, which needed the construction of a quasi-optical band-
rejection filter to reduce incident radiation at the frequency of
this side-band. The data reduction were made in the usual way. A
particular problem comes from the hyperfine splitting of the $C^{17}O$
line which consists of three components with relative intensities
2, 4 and 3 and appreciable frequency separation. In order to compare
the intensities of the $C^{17}O$ lines to those of the $C^{18}O$ line, an
artificial $C^{18}O$ spectrum has been created by splitting each observed
$C^{18}O$ spectrum into three components having the relative weights and
frequency separations of the $C^{17}O$ hyperfine components. Figure 1
presents the $C^{17}O$ observed spectra and the $C^{18}O$ artificial spectra
obtained by Encrenaz et al (1973) and Wannier et al (1976). Table 1
gives the derived $C^{17}O/C^{18}O$ ratios with their uncertainties.
Recently, Gardner and Whiteoak (1976), have detected ^{17}OH in the
Sgr A cloud and find $^{17}OH/^{18}OH = 0.24 \pm 0.02$, in general agreement
with the CO results. This summarizes the present status of our
observational knowledge of the $^{17}O/^{18}O$ interstellar isotopic ratio.

3.DISCUSSION

         For all the sources under consideration, the $^{12}C^{16}O$ line
is at least 6 times stronger than the corresponding $C^{17}O$ and $C^{18}O$
lines, insuring that the line intensity ratios give an accurate
measure of the molecule abundance ratio, and, if chemical fractionation
is not present, of the $^{17}O/^{18}O$ isotopic ratio. The ratio R for the
nine observed clouds may be compared to the terrestrial abundance
ratio $R_{ter} = {}^{17}O/^{18}O = 0.186$ (Wedepohl, 1969). R is generally larger
than R_{ter}, the ratios R/R_{ter} ranging from 0.9 to 1.6. There is some
evidence for source to source variations with a maximum deviation
from the mean of less than 40 per cent. In view of the calibration

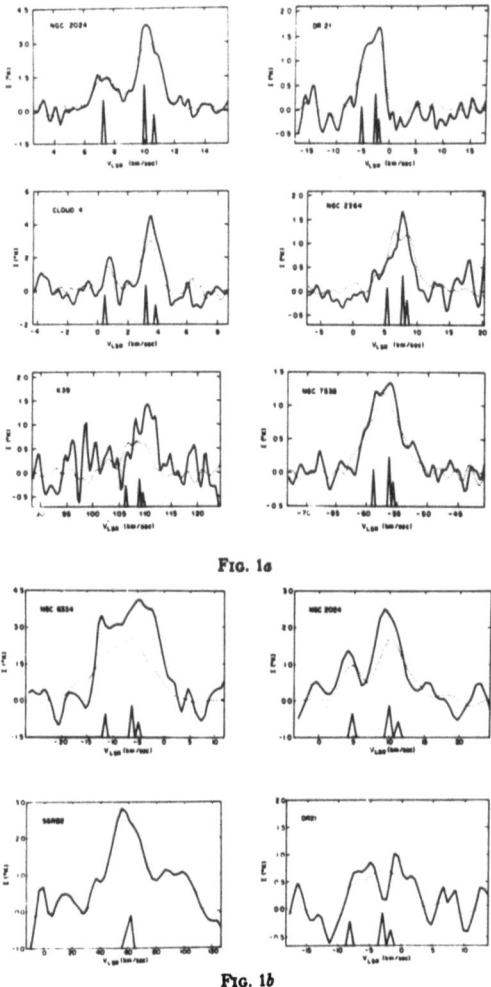

FIG. 1a

FIG. 1b

FIG. 1. We present spectra of $C^{18}O$ (*dotted*) and $C^{17}O$ (*solid*) in eight giant molecule clouds at the $J = 1 \rightarrow J = 0$ rotational transition. The spectra in Fig. 1a are from the NRAO; those in Fig. 1b are from the MWO. The hyperfine splitting of the $J = 1$ level of $C^{17}O$ causes the triple structure which is indicated at the bottom of each spectrum with the velocity resolution used when obtaining the data. The velocity axis corresponds to the most intense line at 112,358.98 MHz. We have modified the $C^{18}O$ spectra by splitting the single spectroscopic feature into three components having the relative weights and velocity separations of the $C^{17}O$ hyperfine components. The $C^{17}O$ spectra have been multiplied by the terrestrial values of $[^{18}O]/[^{17}O] = 5.38$. Thus the two spectra shown for each source should appear identical if the terrestrial isotope ratio applies to the interstellar medium. It is readily seen that the interstellar $[^{17}O]/[^{18}O]$ is generally slightly larger than its terrestrial counterpart. The intensity I corresponds to the brightness temperature above the cosmic background radiation as seen by a perfect antenna and with no atmospheric losses. The differences in the line intensities of DR 21 and NGC 2024 as measured at the two different observatories almost certainly results from the different antenna beamwidths rather than from calibration differences.

TABLE I

Source	Obs.	$R(C^{17}O/C^{18}O) \pm \sigma$ (Terr. = 0.186)	Est. Calib. Uncertainty	$I(CO)/I(C^{18}O)$ $\pm\sigma$	R.A. (1950.0)	Decl. (1950.0)
Orion B.........	NRAO	0.20 ± 0.01	±0.01	13.3 ± 0.6	$05^h39^m12^s$	$-01°55'42''$
Orion B.........	MWO	0.25 ± 0.02	±0.03	13.3 ± 0.6	05 39 12	-01 55 42
Cloud 4(a).......	NRAO	0.22 ± 0.02	±0.01	6.0 ± 0.2	16 23 35	-24 19 00
Cloud 4(b).......	NRAO*	0.25 ± 0.09	±0.03	20 ± 2	16 23 15	-24 19 00
DR 21..........	NRAO	0.22 ± 0.02	±0.01	14.4 ± 0.6	20 37 13	$+42$ 08 51
DR 21..........	MWO	0.19 ± 0.04	±0.02	14.4 ± 0.6	20 37 13	$+42$ 08 51
NGC 7538......	NRAO	0.20 ± 0.02	±0.01	46 ± 5	23 11 17	$+61$ 12 00
NGC 2264.......	NRAO	0.17 ± 0.03	±0.01	12.9 ± 0.7	6 38 25	$+09$ 32 00
K39............	NRAO	0.27 ± 0.05	±0.01	...	18 33 26	-07 15 54
Sgr B2.........	MWO	0.31 ± 0.02	±0.03	10 ± 1	17 44 11	-28 22 30
NGC 6334.......	MWO	0.30 ± 0.02	±0.03	8.1 ± 0.4	17 17 29	-35 44 00
Orion A.........	NRAO*	0.27 ± 0.06	±0.03	61 ± 4	5 32 47	-05 24 30

NOTE.—Along with the ratio $R(C^{17}O/C^{18}O)$ for the nine regions included in our survey we also include the expected statistical uncertainty due to noise, as well as the estimated calibration uncertainty for each determination of R. This last uncertainty is likely to be systematic for each observing run (see text). We note that, in most regions, R is significantly higher than its terrestrial counterpart, though by a factor which never exceeds 1.6. The large values of $I(CO)/I(C^{18}O)$ presented in column (4) demonstrate the unlikelihood that line saturation affects the very weak $C^{18}O$ spectra.
* Encrenaz *et al.* 1973.

Table 1 - Table 1 and Figure 1 come from Wannier et al. (1976)
Co. The University of Chicago Press.

uncertainties, these variations are not very significant. Furthermore, there is no evidence that the $^{17}O/^{18}O$ abundance ratio varies with galactocentric distance.

It appears that the interstellar $^{18}O/^{16}O$ ratio is approximately equal to the solar-system ratio throughout the galactic disk (Wannier, this book), thus the $^{17}O/^{16}O$ ratio is everywhere slightly larger than the solar system ratio. However Whiteoak and Gardner (1976) have given evidences for a twice larger $^{18}O/^{16}O$ ratio in the galactic center (Sgr A and Sgr B2) indicating a possible enrichment in ^{18}O : the $^{17}O/^{16}O$ ratio would thus be accordingly higher in this region of the Galaxy.

The nucleosynthesis of ^{17}O and ^{18}O and the enrichment of the interstellar medium in these isotopes through mass-loss from stars are still complex and poorly-known matters (see Truran, this book). An enrichment in ^{17}O has been seen only in the carbon stars IRC + 10216, and in the red giants Alpha Her and Alpha Sco (Rank et al, 1974 ; Maillard, 1974). The present work provides more observational basis for future theoretical studies.

REFERENCES

Encrenaz, P.J., Wannier, P.G., Jefferts, K.B., Penzias, A.A. and
        Wilson, R.W., 1973, Ap. J. (Letters), 186, L77.

Gardner, S.S., Whiteoak, J.B., 1976, Mon. Not. R. astr. Soc., in press.

Maillard, J.P., 1974, Highlight in Astronomy, Vol. III, Ed. G.
        Contopoulos, D. Reidel, p. 269.

Rank, D.M., Geballe, T.R., Wollman, E.R., 1974, Ap. J. (Letters),
        187, L111.

Wannier, P.G., Lucas, R., Linke, R.A., Encrenaz, P.J., Penzias, A.A.
        and Wilson, R.W., 1976, Ap. J. (Letters), 205, L169.

Wedepohl, K.H., 1969, ed. Handbook of Geochemistry, Springer-Verlag,
        Berlin.

Whiteoak, J.B. and Gardner, S.S., 1975, Proc. Astron. Soc. Australia,
        II, 360.

A MEASUREMENT OF THE INTERSTELLAR [^{12}C/^{13}C] ISOTOPIC ABUNDANCE RATIO TOWARD ζOPH FROM OBSERVATIONS OF THE λ3957 LINE OF CH^{+}

R. Snell, R. Tull, P. Vanden Bout, and S. Vogt
Astronomy Department and McDonald Observatory
University of Texas at Austin

Until recently no visible wavelength line of a secondary isotopic species of any atom or molecule had been detected in the interstellar medium with the sole exception of the λ4232 line of ^{13}CH^{+} which is seen in absorption against ζOph (Bortolot and Thaddeus, 1969; Vanden Bout, 1972). This single line together with rocket obser-vations of the CO ultraviolet bands (Smith and Stecher, 1969) has provided our total knowledge of the relative abundance of ^{13}C in the very local (d < 200 pc) interstellar gas contained in the diffuse clouds typified by the ζOph cloud.

Diffuse clouds are currently of particular interest to those en-gaged in the determination of interstellar isotopic abundances in that these clouds provide a test for the process of chemical fraction-ation of ^{13}C in interstellar CO that has been proposed by Watson, Anicich, and Huntress (1976). If their mechanism operates at all in the diffuse clouds then it must be considered when deriving isotopic abundances from the radio spectral lines observed in the much more dense molecular clouds. Because the dense molecular clouds offer the only potential means for learning relative isotopic abundances throughout the Galaxy, the question of whether there is chemical fractionation is an important one.

Watson et al. state that the "cleanest prediction" they can make is for the diffuse clouds, where

$$[^{13}CO/^{12}CO] \simeq h^{-1}[^{13}C/^{12}C]_{o}.$$

Here $h^{-1} = \exp(\Delta E/k\,T_k)$, where $\Delta E/k = 35K$ is the zero point energy difference between ^{13}CO and ^{12}CO and T_k is the kinetic temperature in the cloud. The ratio $[^{13}C/^{12}C]_{o}$ is the "true" interstellar isotopic abundance ratio. Molecules other than CO, for example CH^{+}, should reflect this ratio. Taking T_k in the range 50-80K for the ζOph cloud we have

$$[^{12}CH^{+}/^{13}CH^{+}] \approx (1.5\text{-}2.0)\ [^{12}CO/^{13}CO].$$

Jean Audouze (ed.), CNO Isotopes in Astrophysics, 85-88. All Rights Reserved.
Copyright © 1977 by D. Reidel Publishing Company, Dordrecht-Holland.

The existing measurements for CH$^+$ are shown in the following table:

Observer	$[^{12}CH^+/^{13}CH^+]$
Bortolot and Thaddeus (1969)	43(+29,-8)
Hobbs (1972)	>52
Vanden Bout (1972)	75 (+25,-15)

The value attributed to Bortolot and Thaddeus is the result of a re-analysis of their data using the velocity dispersion parameter determined by Hobbs (1972) in the curve-of-growth. How significant are these values for the problem at hand? The spectral range scanned by Vanden Bout is too short to allow an accurate determination of the continuum level. A stronger isotopic feature and smaller value for the ratio cannot be ruled out. On the other hand, the two actual detections of the isotopic feature in the table may be of a line that is really a blend. The ζOph cloud has velocity components in atomic lines at -15, -29, and -31.6 km s^{-1}. Thus the $^{13}CH^+$ line from the main -15 km s^{-1} cloud could in principle be blended with $^{12}CH^+$ lines from the other clouds. (The isotopic shift for the λ4232 line is 0.26 Å to the blue.)

In view of these problems in the existing data and the systematic problems of the λ4232 line involving other velocity components we decided to observe the λ3957 line of CH$^+$ toward ζOph. For this line the isotopic shift is 0.44 Å to the red, well away from the potential confusion of other velocity components. Furthermore, the strength of λ3957 to λ4232 is ~2/3 making sure that the $^{12}CH^+$ line is optically thin.

The observations consist of 11 hours of integration on ζOph using the coudé spectrograph of the 2.7 m telescope at McDonald Observatory. A self-scanned silicon diode array (Reticon Corp.) was used as the detector (Vogt and Tull, 1976). An echelle grating was used in 57th order to disperse the spectrum with an 800μ entrance slit and an interference filter to sort grating orders. This provided 1024 simultaneous channels spaced by 0.0158 Å. The spectrograph resolution was 0.13 Å or 8 diode spacings. The data obtained are shown in Figure 1.

The signal strength in the continuum is ~8x10^6 detected photons/channel. Readout noise in the diode array limited the signal-to-noise ratio to ~1500, about half that expected from photon statistics alone. The λ3957 $^{12}CH^+$ is present with a strength of 13.50 ± 0.04 mÅ equivalent width, where the error (and in all other errors quoted) is strictly the 1σ formal error calculated from the rms noise in the data. No systematic errors have been estimated. A feature is easily discerned 0.44 Å to the red of the $^{12}CH^+$ line, precisely where the $^{13}CH^+$ line is expected. The equivalent width of this feature is 0.15 ± 0.04 mÅ. We believe this is the faintest optical interstellar feature ever to be detected.

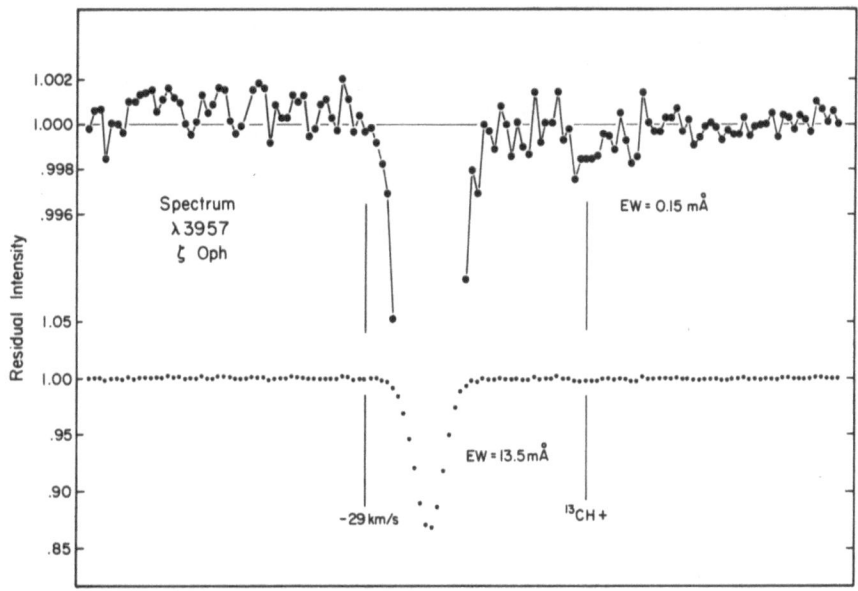

Figure 1.   Spectrum of ζOph in the vicinity of λ3957.

The resulting ratio is [^{13}CH$^+$/^{12}CH$^+$] = 0.011 ± 0.003, or, as it is more conventionally expressed, [^{12}CH$^+$/^{13}CH$^+$] = 90 (+35,-20), a number embarrassingly close to the terrestrial value. Taken by itself this measurement is firm evidence for a large value for the [^{12}CH$^+$/^{13}CH$^+$] ratio in the ζOph cloud. Even the 3σ limits imply a value greater than 50. The ratio [^{12}CO/^{13}CO] = 105 measured by Smith and Stecher (1971) has been revised by them to [^{12}CO/^{13}CO] = 79. This would seem to indicate that chemical fractionation of ^{13}C into ^{13}CO is not taking place in the ζOph cloud. However, the signal-to-noise ratio in these rocket data is less than what might be hoped for and any conclusions on the matter should be held until more recent Copernicus data (Wannier, 1976) has been analysed.

Finally, a warning concerning the optical data must be given. The strength of the ^{13}CH$^+$ feature at λ4232 as seen by Thaddeus, potentially present in the data of Vanden Bout if the continuum were drawn higher, and indicated by recent data we have taken could have been due to other velocity components of the ^{12}CH$^+$ line as was mentioned earlier. These velocity components are not present in the λ3957 data as can be seen by looking in the vicinity of -29 km s^{-1} in Figure 1. Taking the λ4232 data at face value then implies much lower values of [^{12}CH$^+$/^{13}CH$^+$] than 90, for example, the value 43 quoted for the data of Bortolot and Thaddeus. Lacking a clear explanation for the strength of the λ4232 "isotopic feature" at the present time we can only conclude that it is weak evidence for a low value of the [^{12}CH$^+$/^{13}CH$^+$] ratio (the line could at least in principle be a blend with an unknown feature) whereas the λ3957 data are harder evidence for a large ratio.

This work was supported in part by Robert A. Welch Foundation Grant F-623 (P.Vd.B.), NSF Grant AST75-02404 (R.T.), and NASA Grant NGR 44-012-152.

REFERENCES

Bortolot, V.J., and Thaddeus, P.: 1969, Astrophys. J. Letters 155, L17.
Hobbs, L.M.: 1972, Astrophys. J. Letters 175, L39.
Smith, A.M., and Stecher, T.P.: 1971, Astrophys. J. Letters 164, L43.
Vanden Bout, P.A.: 1972, Astrophys. J. Letters 176, L127.
Vogt, S., and Tull, R.: 1976, in preparation.
Wannier, P.: 1976, private communication.
Watson, W.D., Anicich, V.G., and Huntress, W.T.: 1976, Astrophys.J.
    Letters 205, L165.

(^{12}C/^{13}C) RATIOS OBTAINED FROM RADIOASTRONOMICAL MEASUREMENTS

T. L. Wilson and J. H. Bieging
Max-Planck-Institut für Radioastronomie
Bonn, FRG

This paper summarizes the radio astronomical observations of carbon-containing molecules in the interstellar medium. Table 1 contains the (^{12}C/^{13}C) isotope ratios derived from observations reported up to August 1976. Except for the Orion Molecular Cloud, all H$_2$CO ratios were obtained from the 1_{11}-1_{10} absorption line, at 4.8 GHz. The HC$_3$N data of Churchwell et al. ((1976), ref. 7 in Table 1) were measured at 9 GHz. All other isotope results were obtained in the millimeter wavelength region. To minimize the possibility of saturation, the isotope ratios for CO, CS, and HCN were measured using two rarer species, such as ^{13}C^{16}O and ^{12}C^{18}O. Hence the results involve "double ratios", i.e. the product of two isotopic ratios. In Table 1, we have assumed that all isotopic ratios except the (^{12}C/^{13}C) ratio are terrestrial. We have multiplied the observed intensity ratio (^{12}C^{18}O)/(^{13}C^{16}O) by the terrestrial (^{16}O/^{18}O) ratio, 489, the (^{12}C^{34}S)/(^{13}C^{32}S) value by the terrestrial (^{32}S/^{34}S) ratio, 23, and (H^{12}C^{15}N)/(H^{13}C^{14}N) by the terrestrial (^{14}N/^{15}N) ratio, 269. Wannier et al. (1976) and R. W. Wilson et al. (1976) have argued that the (^{16}O/^{18}O) and (^{32}S/^{34}S) ratios are close to terrestrial, but Linke et al. (1976) conclude that ^{14}N is enriched in the galactic center region. Clearly, if both isotope ratios are different from terrestrial, it is difficult to draw any conculsions about the (^{12}C/^{13}C) ratio from CO, CS or HCN double ratios.

The ratios obtained from the 4.8 GHz H$_2$CO absorption lines toward strong discrete sources (refs. 1 and 2 in Table 1) were calculated from

$$R = 0.95 \; \frac{\int \tau(H_2{}^{12}CO)\,dV}{\int \tau(H_2{}^{13}CO)\,dV} = \frac{\left(\dfrac{N_{1_{11}}(H_2{}^{12}CO)}{T_{ex}(H_2{}^{12}CO)} \right)}{\left(\dfrac{N_{1_{11}}(H_2{}^{13}CO)}{T_{ex}(H_2{}^{13}CO)} \right)} \tag{1}$$

where $\tau = \ln(1 - T_{Line}/T_{continuum})$. The H$_2$CO results in Table 1 were obtained

Jean Audouze (ed.), CNO Isotopes in Astrophysics, 89-93. All Rights Reserved.
Copyright © 1977 by D. Reidel Publishing Company, Dordrecht-Holland.

Table 1

(¹²C/¹³C) RATIOS FOR MOLECULAR CLOUDS

(1)	(2)	(3)	(4)	(5)	(6)	(7)	(8)	(9)
Cloud	V_{LSR} (km s^{-1})	$\dfrac{H_2^{12}CO}{H_2^{13}CO}$	$\dfrac{^{12}C^{18}O}{^{13}C^{16}O}\cdot 489$ (All data from (4))	$\dfrac{^{12}C^{34}S}{^{13}C^{32}S}\cdot 23$	$\dfrac{H^{12}C^{15}N}{H^{13}C^{14}N}\cdot 269$	HCCCN $\dfrac{HCCCN}{H^{13}C^{12}C^{12}CN}$	$\dfrac{HCCCN}{H^{12}C^{13}C^{12}CN}$	$\dfrac{HCCCN}{H^{12}C^{12}C^{13}CN}$
G350.1+0.1	-69.4[a]	---	51±14 (r_o=3.5) [b]	---	---	---	---	---
NGC 6334	- 4.7	---	49±14 (r_o=3.5)	---	34±7 (9)	---	---	---
ρ Oph Dark Dust Cloud	+ 2.8	---	44± 6[c]	---	---	---	---	---
Sgr A "0 km s⁻¹ Feature"	- 4.0	49± 4 (1)	---	---	---	---	---	---
Sgr A "+40 km s⁻¹ Feature"	+42.5	>13 (1)	---	---	2±5 (9)	---	---	---
Sgr B2	+62.5	>10[d] (1)	43± 4 (r_o=3.5)	38± 4 (5)	6±5 (9)	[21± 3 (7) / >20 (6)]	56±10 (7) / >20 (6)	56±10 (7) / >16 (6)
M 8	+11.3[a]	---	26± 9 (r_o=4.5)	---	---	---	---	---
W 31	+ 8.9	38± 3 (1)	56± 9 (r_o=2.5)	---	---	---	---	---
W 33	+33.8	22± 1 (1)	---	---	---	---	---	---
M 17	+20.6[a]	>70 (2)	52± 4 (r_o=2.5)	37± 2 (5)	---	---	---	---
K 39	+112.6[a]	---	105± 2 (r_o=1.5)	---	---	---	---	---
W 43	+91.2	31± 2 (1)	---	---	---	---	---	---
W 49A	+14.3	40± 5 (1)	31±11 (r_o=1.5)	---	---	---	---	---
G49.4-0.3	+62.5	73±17 (1)	---	---	---	---	---	---
W 51[e]	+66.4	37± 2 (1)	63± 4 (r_o=2.5)	60±11 (5)	[48±13 (4) / 56± 5 (9)]	---	---	---
DR 21	- 2.6	49± 4 (1)	29± 7 (r_o=5.5)	---	---	---	---	---
NGC 7538	-55.5[a]	---	38± 6 (r_o=4.5)	---	44± 8 (9)	---	---	---
Cassiopeia A	- 0.2 / - 1.6]	>90 (2)	---	---	---	---	---	---
	-39.6	>44 (3)	---	---	---	---	---	---
	-46.6	>63 (3)	---	---	---	---	---	---
W 3	-39.5	[79±19 (1) / 62±15 (2)]	---	---	---	---	---	---
Taurus Dark Dust Cloud	6.6	33[g] (3)	---	---	---	---	---	---
Orion Dark Dust Cloud	+ 8.4	>28[g] (3)	---	---	---	---	---	---
Orion Molecular Cloud	+ 9.3[a]	~77 (8) / 32± 7 (4)	37± 2 (r_o=4.5)	60± 3 (5)	[75±16 (4) / 67± 5 (9)]	---	---	---
NGC 2024	+ 9.3	[43± 5 (1) / 55± 6 (2)]	19± 6[f] (r_o=4.5)	---	---	---	---	---
NGC 2264 (Dark Dust Cloud)	+ 6.4	>32[g] (3)	40±12 (r_o=4.5)	62± 8 (5)	---	---	---	---
L 134 (Dark Dust Cloud)	+ 3.3	>31[g] (3)	---	---	---	---	---	---
L 134 N (Dark Dust Cloud)	+ 2.6	20[g] (3)	---	---	---	---	---	---

(1) T.L. Wilson, Bieging, Downes, Gardner (1976)
(2) Matsakis, Chui, Goldsmith, Townes (1976)
(3) Evans, Zuckerman, Morris, Sato (1975)
(4) Wannier, Penzias, Linke, Wilson,R.W. (1976)
(5) R.W. Wilson, Penzias, Wannier, Linke (1976)
(6) Morris, Turner, Palmer, Zuckerman (1976)
(7) Churchwell, Walmsley, Winnewisser (1976)
(8) Kutner, Evans, Tucker (1976)
(9) Linke, Goldsmith, Wannier, R.W. Wilson, Penzias (1976)

All errors are one standard deviation.
a) Velocity of $^{12}C^{18}O$, otherwise velocity of $H_2^{13}CO$.
b) r_o is the ratio of the temperatures of $^{12}C^{16}O$ to $^{13}C^{16}O$. The comment indicates that the ratio of ($^{12}C^{18}O/^{13}C^{16}O$) has been formed in the portion of the profiles where r_o is larger than the value given, i.e. in the line wings.
c) Data from Encrenaz(1974), reduced using method of (4).
d) Ratio of($H_2^{12}C^{18}O/H_2^{13}C^{16}O$) gives 0.1±0.01 (Gardner et al.1971) in agreement with ($^{12}C^{18}O/^{13}C^{16}O$) ratio of Wannier et al. (1976).
e) The largest ^{13}CO emission occurs at +56 km s⁻¹.
f) The ratio ($^{12}C^{18}O/^{13}C^{18}O$)=48±11 (4).
g) Ratio of the equivalent widths.

with telescope beamwidths $\geq 2.6'$, so structure in the H_2CO on a smaller angular scale may affect the (^{12}C/^{13}C) ratios. However, from a comparison with the interferometer observations of the 4.8 GHz line of $H_2^{12}CO$ (Fomalont and Weliachew 1973), we find that corrections to R because of an underestimation of τ should be less than 20% for sources outside the galactic center. The isotope ratios for Dark Dust Clouds (ref. 3) were taken from the ratios of the equivalent widths. For these clouds, the optical depth in the $H_2^{12}CO$ line at 4.8 GHz is usually ≥ 1 (Heiles 1973), so that the ratios are lower limits.

The derived ratio R in Equation (1) depends on both the column densities and the excitation temperatures of the observed lines. Transitions (at 140 and 150 GHz) from the ground state of formaldehyde to the next rotational state may be optically thick for $H_2^{12}CO$, but not for $H_2^{13}CO$. As a consequence, the ground state of $H_2^{12}CO$ may be underpopulated relative to $H_2^{13}CO$. At the same time, the excitation temperature of the 6cm transition in $H_2^{12}CO$ may be raised relative to that in $H_2^{13}CO$. Thus, except for the most tenuous clouds, corrections for photon trapping in H_2CO will tend to raise the observed ratio R. Hence the values in column 3 of Table 1 are lower limits, if the only systematic effects are caused by differences in the 6cm excitation temperatures and rotational populations for $H_2^{12}CO$ and $H_2^{13}CO$.

The CO isotope ratios (column 4 of Table 1) are perhaps the most difficult to interpret because the optical depth at the center of the $^{13}C^{16}O$ line might be ~ 1. To avoid such saturation problems, Wannier et al. (1976) have used data in the wings of the $^{13}C^{16}O$ and $^{12}C^{18}O$ lines, where the optical depth is expected to be lower. Wannier et al. have used the quantitative criterion that the $^{13}C^{16}O$ line is optically thin when r_o, the ratio of the antenna temperatures of $^{12}C^{16}O$ to $^{13}C^{16}O$ is larger than 3.5. We have included results for W31, M17, K39, W49A and W51, for which r_o is less than 3.5. If the $^{13}C^{16}O$ line is optically thick, the observed R will be too large, and the corrections will lower the ratio, if the excitation temperature of $^{12}C^{18}O$ is equal to that of $^{13}C^{16}O$. The optical depths of CS and HCN are lower than the optical depth of the CO, and the entire line profiles were used for the calculation of the isotope ratios in column 5 and 6. The ratios taken from references 4 and 5 were obtained by weighting the weaker line by the intensity of the stronger line. (Note that if the isotopic ratio is not constant over a given line profile, this procedure will yield a ratio weighted toward the value of R where the line intensity is greatest.) R. W. Wilson et al. (1976) argue that this reduction method does not have a large effect on the CS ratios.

The average value of the isotope ratio for each of the species measured in more than one source are given below:

AVERAGE (^{12}C/^{13}C) RATIOS OBTAINED FROM TABLE 1					
molecule	4.8 GHz H_2CO	CO ($r_o \geq 3.5$)	CO (all r_o)	CS	HCN
(^{12}C/^{13}C) ratio	46±16	38±10	46±20	51±13	35±26
number of sources in average	10	10	15	5	6

The average values agree very well, but the (^{12}C/^{13}C) ratios obtained from CO, CS and HCN were calculated assuming terrestrial (^{16}O/^{18}O), (^{32}S/^{34}S) and (^{14}N/^{15}N) ratios, respectively. Furthermore, the velocities of the deepest H$_2$CO and ^{13}CO features differ for some clouds; for example, for W51, this velocity difference is 10 km s^{-1}. A comparison of the ratios from different molecules for a given source shows a wide scatter of values (cf. the various results obtained for the Orion Molecular Cloud in Table 1). We will consider the data for six molecular clouds.

The "0 km s^{-1} Feature" toward Sgr A is probably a cloud in the Sagittarius spiral arm, about 2 kpc from the sun. Unlike most of the other H$_2$CO clouds, this object and the clouds toward Cas A are not close to HII regions. We believe that the optical depths of the H$_2$^{12}CO lines are low, so corrections for photon trapping and optical depth in H$_2$^{12}CO are small. The limit reported for the Cassiopeia clouds near 0 km s^{-1} is $\geq$ 89 (ref. 2, Table 1), while the ratio for the 0 km s^{-1} cloud toward Sgr A is 49±4. The large difference in these ratios suggests possible cloud-to-cloud variations. The uncertainty in the limit for Cas A is large, however, and further measurements of the H$_2$CO toward Cas A might also give a lower value.

The ratios for the molecular clouds near Sgr A (the "+ 40 km s^{-1} Feature") and Sgr B2 vary between 2 and 43. The very low R values obtained from HCN (column 6 of Table 1) have been interpreted by Linke et al. (1976) as evidence for the enrichment of ^{14}N; hence the (^{12}C/^{13}C) ratios obtained from HCN are uncertain. For Sgr B2, the CO double ratio varies from 43 (at r_o=3.5, given in Table 1) to 26±13 (at r_o=6.5). The value of 43 agrees with the double ratio obtained from H$_2$^{13}CO and H$_2$C^{18}O (Gardner et al. 1971). The (^{12}C/^{13}C) ratio in column 3 of Table 1, 10, is certainly a lower limit. Fomalont and Weliachew (1973), from an estimate of the optical depth of the H$_2$^{12}CO line, corrected the ratio to $\gtrsim$ 20. If however, we form the (^{12}C/^{13}C) ratio in the wings of the profile, the ratio does not rise but rather remains 10 (Bieging, Downes, Martin, Wilson, in prep.). Finally, the HC$_3$N data of Churchwell et al. (1976), in columns 7 to 9 of Table 1, yield a different ratio for two of the three ^{13}C substitutions. The HC$_3$N line observed at 9 GHz is definitely masing, and it is possible that the different isotope ratios are due to differences in the excitation temperatures, but the effect might also be caused by chemical fractionation (see e.g. Watson et al. 1976). The (^{12}C/^{13}C) ratio for Sgr A and Sgr B2 probably lies between 10 and 40, that is, somewhat lower than the ratio for sources outside the galactic center.

The (^{12}C/^{13}C) ratio obtained from H$_2$CO for the molecular cloud near NGC 2024 is nearly twice the value obtained from the CO double ratio, assuming a terrestrial oxygen abundance. A direct determination of the isotope ratio from the ^{13}C^{18}O line (see footnote (f) of Table 1) agrees with the H$_2$CO determination. Although it is possible that ^{18}O is a factor of two underabundant, the shapes of the ^{13}C^{18}O and ^{12}C^{18}O lines differ, and the double ratio obtained from the CO might be unreliable in this case.

For the Orion Molecular Cloud, the (^{12}C/^{13}C) ratios from the millimeter lines of H$_2$CO differ by a factor of 2 (refs. 4 and 8 in Table 1).

Wannier et al. made no correction for the optical depth of the $H_2^{12}CO$ line, while Kutner et al. have attempted an elaborate correction process. A large part of the difference is the influence of noise on the $H_2^{13}CO$ profiles.

CONCLUSIONS

The (^{12}C/^{13}C) isotope ratios determined from H_2CO, CO, CS, and HCN in a number of galactic sources show that the ratios are generally lower than terrestrial. For individual sources, the ratios determined from different molecules show a large scatter. It is possible that chemical fractionation has some effect, but this process is unlikely to account for all of the observed variations. Because many of the isotope measurements involve the product of two isotope ratios, it is difficult to determine how much of the change in the ratios is due to a change in (^{12}C/^{13}C) alone. The reality of source-to-source variations is not clear. The H_2CO results of T. L. Wilson et al. (1976) for 8 of 10 clouds and the CO results of Wannier et al. (1976) for all clouds showed no variations at the two standard deviation level. The H_2CO data of Matsakis et al. (1976), the CS ratios of R.W. Wilson et al. (1976) and the HCN survey of Linke et al. (1976), however, support at least some variations. It should be emphasized that the ratios obtained are based on measurements of only one transition from each molecular species and the total column density of the species is obtained from a number of assumptions, which may not be valid in all cases.

REFERENCES

Churchwell, E.B., Walmsley, C.M., Winnewisser, G., 1976, Astron. Astrophys., in press

Evans, N.J.II, Zuckerman, B., Morris, G., Sato, T., 1975, Astrophys. J. 196, 433

Fomalont, E.B., Weliachew, L., 1973, Astrophys. J. 181, 781

Gardner, F.F., Ribes, J.C., Cooper, B.F.C., 1971, Astrophys. Lett. 9, 181

Heiles, C., 1973, Astrophys. J. 183, 441

Kutner, M.L., Evans, N.J.II, Tucker, K.D., 1976, Astrophys. J., in press

Linke, R.A., Goldsmith, P.F., Wannier, P.G., Wilson, R.W., Penzias, A.A., 1976, Astrophys. J., in press

Matsakis, D.N., Chui, M.F., Goldsmith, P.F., Townes, C.H., 1976, Astrophys. J. 206, L63

Morris, M., Turner, B.E., Palmer, P., Zuckerman, B., 1976, Astrophys. J. 205, 82

Wannier, P.G., Panzias, A.A., Linke, R.A., Wilson, R.W., 1976, Astrophys. J. 204, 26

Watson, W.D., Anichich, V.G., Huntress, W.T., 1976, Astrophys. J. 205, L165

Wilson, R.W., Penzias, A.A., Wannier, P.G., Linke, R.A., 1976, Astrophys. J. 204, L135

Wilson, T.L., Bieging, J., Downes, D., Gardner, F.F., 1976, Astron. Astrophys. 51, 303.

ISOTOPE RATIOS AND CHEMICAL FRACTIONATION OF CO IN LYNDS 134

R.L. Dickman
Aerospace Corporation

W.D. Langer
Goddard Institute for Space Studies NASA

W.H. McCutcheon and W.L.H. Shuter
University of British Columbia

Mahoney, McCutcheon and Shuter (1976) reported observations of the $J = 1 \rightarrow 0$ transition of three isotopes of CO in the dust cloud Lynds 134 using the 4.6 m telescope at Aerospace Corporation. Here we discuss a new observation of $^{12}C^{17}O$ and consider further the question of the ratio $^{13}C^{16}O/^{12}C^{18}O$ across the dust cloud.

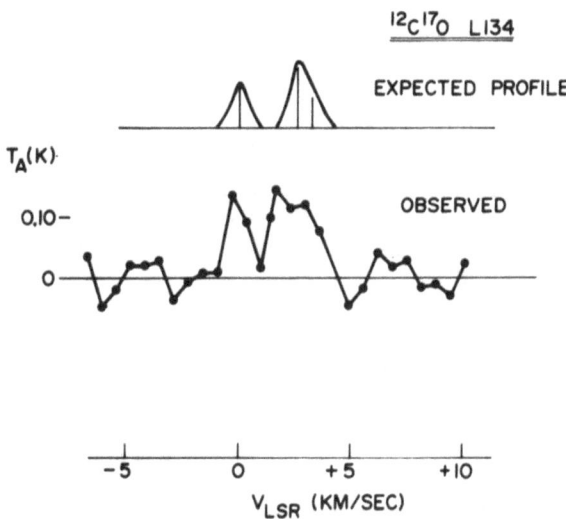

Figure 1. Observation of $^{12}C^{17}O$ in Lynds 134 at $\alpha(1950)$ 15^h 51^m 00^s, $\delta(1950)$ - $4°30'00''$. The radial velocity scale applies to the most intense hyperfine component.

Jean Audouze (ed.), CNO Isotopes in Astrophysics, 95-98. All Rights Reserved.
Copyright © 1977 by D. Reidel Publishing Company, Dordrecht-Holland.

We obtain a ratio of column densities $^{12}C^{17}O/^{12}C^{18}O = 0.28 \pm 0.07$
which differs from the terrestrial value of 0.18 but is in agreement
with the results of Wannier et al. (1976) and Gardner and Whiteoak (1976)
that the $^{17}O/^{18}O$ ratio in interstellar space is somewhat larger than
the terrestrial ratio.

We reconsider the question of ratios $^{13}C^{16}O/^{12}C^{18}O$ in Lynds 134
and the nature of the problem is demonstrated in Figure 2. Here we
show spectra of CO isotopes at two positions in Lynds 134 which are
separated by about 3.75 arcmin or 1.5 beamwidths of the antenna. It is
quite apparent that the ratio $^{13}C^{16}O/^{12}C^{18}O$ is very different in the
two positions.

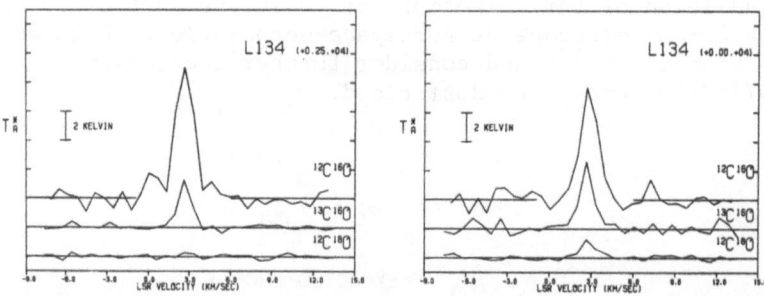

Figure 2.   The left hand diagram shows plots of spectra of $^{12}C^{16}O$ (top)
$^{13}C^{16}O$ (center), $^{12}C^{18}O$ (bottom) at α(1950) 15^h 51^m 15^s, δ(1950)
$- 4°26'00"$, while the right hand diagram shows similar plots at α(1950)
15^h 51^m 00^s, δ(1950) $- 4°26'00"$.

In Figure 3 is shown the ratios of column densities $^{13}C^{16}O/^{12}C^{18}O$
calculated on the assumption of LTE. It is found that at six positions
in the most obscured part of the cloud the mean value of this ratio is
4.8 and that there is not much scatter about this mean. This should be
compared with the terrestrial ratio of 5.6. At positions outside the
core of the cloud the ratio is much larger and variable. In fact we
find a rough inverse correlation between the ratio and the visual
extinction Av. At low values of Av the ratio is high and at high values
of Av the ratio tends to a uniform value of about 4.8.

We initially thought that this result could be explained by depar-
tures from LTE, but a full non LTE analysis was only able to produce a
rather small variability in the ratio $^{13}C^{16}O/^{12}C^{18}O$, certainly less
than a factor of 2. As a result we now believe that we are seeing
chemical fractionation of CO, and that in the outer parts of Lynds 134
there is considerable enrichment of $^{13}C^{16}O$ by the reaction described
by Watson et al. (1976):

$$^{13}C^{+} + {}^{12}CO \rightleftharpoons {}^{12}C^{+} + {}^{13}CO + \Delta E.$$

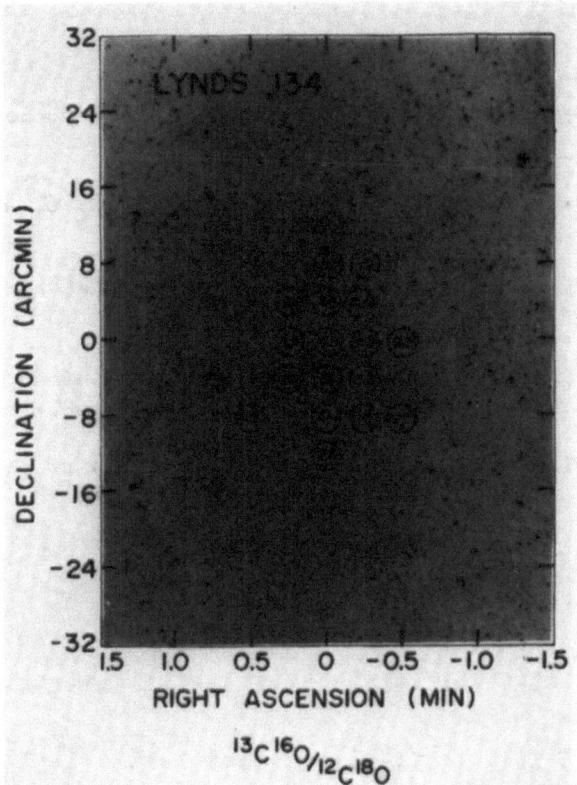

Figure 3.   Shown are the ratios of $^{13}C^{16}O/^{12}C^{18}O$ column densities calculated assuming LTE. The circles represent the positions observed
and are the size of the antenna beam (FWHM), and they are superposed
on the Palomar Sky Survey red plate. The diagram is centered at α(1950)
$15^h 51^m 00^s$, δ(1950) $-$ 4°30'00".

     Lynds 134 appears to be a dust cloud with a larger than average
incident uv flux which destroys some of the hydrogen molecules, and
this leads to the detectability of 21 cm absorption. It therefore
would be expected to have a higher fractional abundance of ions than the
average cloud, and therefore a greater possibility for chemical fractionation. Some support for this view is given by recent observations
of DCO^{+} by Hollis et al. (1976) who find the highest ratio of
$DCO^{+}/HCO^{+} \sim 1$ in Lynds 134.

     Finally our ratio of $^{13}C^{16}O/^{12}C^{18}O$ = 4.8 in the cloud core, where
we do not believe there is appreciable fractionation, implies a
$^{12}C/^{13}C$ ratio $\sim$ 104 which is in approximate agreement with the terrestrial
ratio of 89. This result is in approximat agreement with the results
of Vanden Bout obtained from optical observations of $^{12}CH^{+}/^{13}CH^{+}$ but
is not in agreement with the results of Wannier who from observations

of several molecules suggests the ratio of $^{12}C/^{13}C$ in the interstellar
medium is $\sim$ 40.

REFERENCES

Gardner, F.F., and Whiteoak, J.B.: 1976, Mon. Not. R. astr. Soc. 176,
    53^P.

Hollis, J.M., Snyder, L.E., Lovas, F.J., and Buhl, D.: 1976, Astrophys.
    J. Lett. In press.

Mahoney, M.J., McCutcheon, W.H., and Shuter, W.L.H.: 1976, Astron. J.
    In press.

Wannier, P.G., Lucas, R., Linke, R.A., Encrenaz, P.J., Penzias, A.A.,
    and Wilson, R.W.: 1976, Astrophys. J. Lett. 205, L169.

Watson, W.D., Anicich, V.G., and Huntress, W.T.: 1976, Astrophys. J.
    Lett. 205, L165.

OBSERVATIONAL EVIDENCE FOR CHEMICAL FRACTIONATION IN SGR B2: THE ^{13}C-ISOTOPES OF CYANOACETYLENE

E. Churchwell, C. M. Walmsley and G. Winnewisser
Max-Planck-Institut für Radioastronomie, Bonn,
Germany, and Institut für Physikalische Chemie,
Justus-Liebig-Universität, Giessen, Germany.

INTRODUCTION

To derive abundance ratios from measured line intensities several fundamental assumptions are usually made. The two most important of these are that:
1) the opacity is small in the lines of the ^{12}C-species and that trapping in higher rotational levels does not significantly alter the populations of the measured transition (i.e. T_{ex} is the same for the ^{12}C- and ^{13}C-species); and
2) no chemical fractionation takes place (i.e. all molecules have the same concentration of ^{13}C-substituted species relative to the parent molecule).

HC_3N ($\equiv$ $H^{12}C^{12}C^{12}C^{14}N$) is an especially appropriate tool for studying the ^{12}C/^{13}C abundance problem in Sgr B2. Firstly, its rotational transitions are intense enough that ^{13}C-substitutions can be observed in all three different locations in the molecule. Secondly, substantial observational evidence indicates that the opacity in the $J = 1 - 0$ line is small so that the measured intensity ratio of the ^{12}C-species with any one of its three ^{13}C-species should give an accurate approximation of the ^{12}C/^{13}C ratio if T_{ex} is the same for all species.

Gardner and Winnewisser (1975) detected two ^{13}C-species ($HC^{13}CCN$ and $HCC^{13}CN$)* in Sgr B2. They found that both ^{13}C-species have the same intensity within the errors and that the intensity ratio

$$\frac{I(HC_3N)}{I(HC^{13}CCN)} \simeq \frac{I(HC_3N)}{I(HCC^{13}CN)} \simeq 36 \pm 5.$$

We report the results of two independent observations of all three single ^{13}C-substituted species and the ^{12}C-species of cyanoacetylene in Sgr B2 and briefly discuss the implications of these results with regard to interstellar isotopic abundance determinations and chemical fractionation.

---

* In the rest of this paper, we will denote ^{12}C and ^{14}N by C and N respectively.

Jean Audouze (ed.), CNO Isotopes in Astrophysics, 99-103. All Rights Reserved.
Copyright © 1977 by D. Reidel Publishing Company, Dordrecht-Holland.

OBSERVATIONS

    The observations reported here were made with the 100-m telescope
in Effelsberg, W. Germany. In the frequency range 8.8 - 9.1 GHz the half-
power beamwidth (HPBW), aperture and beam efficiencies were $\sim$90" arc,
0.46 and 0.65 respectively. The spectrometer was a 384 channel autocorre-
lator used in the dual receiver mode (i.e. two independent 192 channel
spectrometers) with total bandwidths of 10 MHz and a resolution of
$\sim$63 kHz (i.e. 2.1 km s^{-1}).

    All spectra were independently observed in two different observing
periods, each with a different receiver system: the first session was in
Oct. - Nov., 1975 with a cooled dual channel paramp whose system temperatures (T_{sys}) were $\sim$90K and $\sim$100K on cold sky; the second session was in
May, 1976 with a single channel cooled paramp ($T_{sys} \approx$ 75K on cold sky).
The line intensities and shapes were the same within the noise in both
sessions.

    The observed spectra with either a linear or quadratic baseline re-
moved are shown in figures 1 - 3. In the bandpass of the H^{13}CCCN spectrum
(Figure 2) lies the $5_{23} - 6_{16}$, J = 11/2 - 13/2, F = 9/2 - 11/2 transition
of NH$_2$ at 8819.81 MHz (Cook et al. 1976). From our data we can set an
upper-limit of 0.03 K on the peak line temperature for this transition.
The spectrum shown in Figure 3 is complicated by the fact that the velocity components and hyperfine components of the two ^{13}C substituted species overlap somewhat. From fitting a pattern of gaussian functions (in
which the relative separations were held fixed to the measured laboratory
separations of the quadrupole components and the line widths were held
fixed to that of the unblended F = 2 - 1 component of HCC^{13}CN) and from
eye estimates one sees that the intensities of the HC^{13}CCN and HCC^{13}CN
species are the same within the noise, a result found also by Gardner
and Winnewisser (1975). Integration over the F = 1 - 1 and F = 2 - 1
components of each species indicates however, that the intensity of the
H^{13}CCCN species is somewhat greater than the combined intensities of the
HC^{13}CCN and the HCC^{13}CN species. The observed line parameters and inte-
grated intensities are given in table 1.

    The velocities of all isotopic substitutions are in excellent agree-
ment and are $\sim$64 km s^{-1}. There is a discrepancy in the measured widths in
the sense that relative to those of HC$_3$N the H^{13}CCCN lines are $\sim$3 km s^{-1}
wider and the HC^{13}CCN and HCC^{13}CN lines are $\sim$4 km s^{-1} narrower. This difference is qualitatively confirmed by comparison with our 1975 data.
Within the uncertainties the hyperfine intensity ratios (F = 1 - 1 /
F = 2 - 1) in the ^{13}C species are consistent with the optically thin LTE
value of 0.6, whereas that of the HC$_3$N species is $\sim$0.39.

    Regardless of how the ratios are calculated, it is apparent that the
intensity of the H^{13}CCCN species is a factor of two or greater than that
of the other two species, HC^{13}CCN and HCC^{13}CN. The observational errors
are much smaller than this difference and the result has been verified by
two independent observations made at different times and with different
receiver systems.

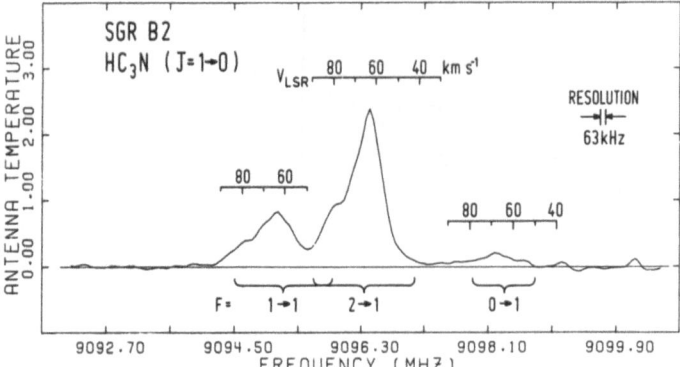

FIGURE 1: THE OBSERVED $J = 1-0$ ROTATIONAL TRANSITION OF HCCCN TOWARD SGR B2. A VELOCITY
SCALE HAS BEEN DRAWN ABOVE EACH HYPERFINE COMPONENT. THE HYPERFINE COMPONENTS ARE
DESIGNATED BELOW EACH LINE. THE FREQUENCY SCALE IS FOR THE REST FRAME OF THE OB-
SERVER. A LINEAR BASELINE HAS BEEN FITTED.

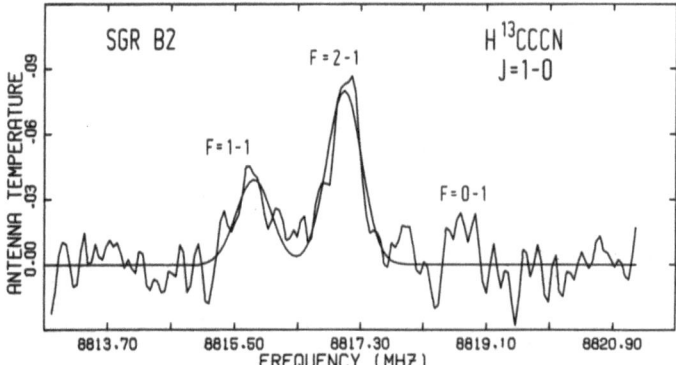

FIGURE 2: THE OBSERVED $J = 1-0$ TRANSITION OF H^{13}CCCN WITH A CUBIC BASELINE REMOVED AND A
GAUSSIAN FUNCTION FITTED TO THE $F = 1-1$ AND $F = 2-1$ COMPONENTS. THE FREQUENCY
SCALE IS THAT OF THE MOLECULAR REST FRAME.

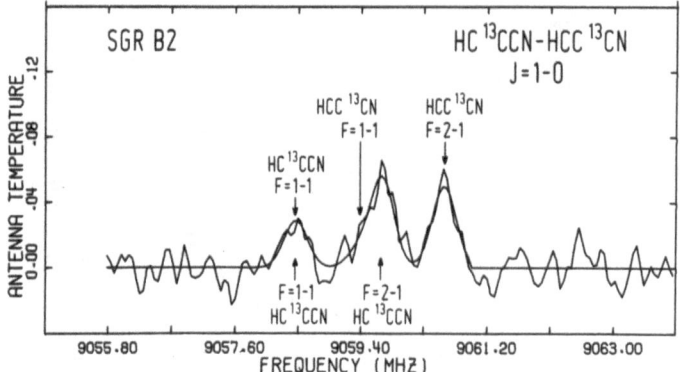

FIGURE 3: THE OBSERVED $J = 1-0$ TRANSITION OF THE BLENDED HC^{13}CCN AND HCC^{13}CN SPECIES WITH
A QUADRATIC BASELINE REMOVED AND A MULTIPLE GAUSSIAN FUNCTION FITTED TO THE OBSERVED
SPECTRUM (SEE TEXT). THE FREQUENCY SCALE IS IN THE REST FRAME OF THE MOLECULE.

Table 1: Measured Line Parameters

Isotope	Hyperfine Component F'-F''	T_{AL} K	Rel. Int.	$\Delta V^{2)}$ km s^{-1}	V_{LSR} km s^{-1}	$I_L^{1)}$ K kHz	$\dfrac{I_{12}}{I_{13}}$
HCCCN	1 → 1	0.82 ± .01	27	17.2 ± 1.0	64.1 ± .5		
	2 → 1	2.11 ± .01	68	17.2 ± 1.0	64.1 ± .5	} 1656 ± 10	
	0 → 1	0.16 ± .01	5	17.2 ± 1.0			
H^{13}CCCN	1 → 1	0.039 ± .01	27	20.4 ± 1.0	64.6 ± 1.0		
	2 → 1	0.080 ± .01	55	20.4 ± 1.0	64.6 ± 1.0	} 80.2 ± 8.0	20.6
	0 → 1	∼0.025 ± .01	17				
HC^{13}CCN	1 → 1	0.025 ± .01		13.4 ± 2.0	64.0 ± 1.5		(56.4)$^{3)}$
	2 → 1	0.055 ± .01		13.4 ± 2.0	64.0 ± 1.5	58.7 ± 8.0	28.2
HCC^{13}CN	1 → 1	(blended)					(56.4)$^{3)}$
	2 → 1	0.050 ± .01		13.4 ± 2.0	64.0 ± 1.5		

Notes:

1) $I_L \equiv \int_{\nu_1}^{\nu_2} T_L \, d\nu$ where the limits span the F = 1 - 1 and F = 2 - 1 transitions.

2) ΔV = full width at half maximum intensity.

3) These values were derived by assuming that the intensities of the HC^{13}CCN and HCC^{13}CN species are equal.

DISCUSSION

The relative intensities of the $J = 1 - 0$ transition in HC_3N and its ^{13}C isotopically substituted species represent the true relative concentrations, if the lines are optically thin and if their excitation temperatures are the same. Present observational evidence (Morris et al. 1976, and our own extensive HC_3N data and analysis, in preparation) suggests that $T_{ex} < 0$, $|T_{ex}| \ll T_c$ and $|\tau| \lesssim 0.2$ at the center of the $F = 2 - 1$ transition of HC_3N.

The next question is whether $T_{ex}(12) \simeq T_{ex}(13)$. The fact that the hyperfine ratios in the ^{13}C-species are consistent with LTE values, within the noise, whereas those of the ^{12}C-species are clearly not, is a direct indication that $T_{ex}(12) \neq T_{ex}(13)$. Our statistical equilibrium calculations show that $T_{ex}(12)$ is sensitive to trapping which, at least to a minor extent must take place in the higher rotational transitions. We therefore conculde that our measured intensity ratios are probably lower limits to the real abundance ratio of HC_3N to its three ^{13}C species. It is improbable that the excitation temperatures of the three ^{13}C-species are different, although we note that a small change in excitation temperature could, for inverted levels, explain the observed difference in the intensities of the ^{13}C-species. It is unlikely that the relative strengths of the three ^{13}C-species are influenced by trapping. For this to be the case, one would have to have large optical depths in the $J = 1 - 0$ transition of $H^{12}C_3N$ and hence even larger optical depths in higher J transitions. The apparent optical depths of the ^{13}C-species are ~ 0.01. If this is the actual optical depth, trapping can not significantly influence the excitation temperatures of the ^{13}C-species of HC_3N in Sgr B2.

CONCLUSION

We believe that chemical fractionation of ^{13}C is taking place in the Sgr B2 cloud causing $H^{13}CCCN$ to be more than twice as abundant as the other two ^{13}C-species. It is unlikely that this is an excitation effect, although verification by observing a higher J transition would be worthwhile and should be done. If fractionation is taking place in HC_3N, $^{12}C/^{13}C$ abundance measurements from other molecules may be similarly affected and should be regarded with suspicion.

REFERENCES

Cook, J. M., Hills, G. W., Curl, R. F. Jr. 1976, Astrophys. J. 207, L135.

Gardner, F. F., Winnewisser, G. 1975, Astrophys. J. (Letters), 197, L73.

Morris, M., Turner, B. E., Palmer, P., Zuckerman, B. 1976, Astrophys. J. 205, 82.

ISOTOPE FRACTIONATION IN INTERSTELLAR MOLECULES[*]

William D. Watson[**]
Departments of Physics and Astronomy
University of Illinois

ABSTRACT

Fractionation of carbon isotopes in interstellar clouds is treated in some detail. Acting together, the exchange reaction $^{13}C^+ + ^{12}CO \rightleftarrows ^{12}C^+ + ^{13}CO + \Delta E$ and the high vapor pressure of CO may cause a preferential loss of ^{12}C from the gas due to freezing onto dust grains. The $^{13}C/^{12}C$ ratio for the carbon in the gas is then raised by factors ~ 2 to 4. Fractionation of deuterium is also discussed since the proposed process is similar to the carbon isotope exchange and here there is no question that fractionation is a major effect. Important information about physical conditions in dense clouds -- the fractional electron density and CO abundance -- can be derived from the fractionation process producing the enhanced DCO^+/HCO^+ ratio. Possible fractionation reactions for isotopes of other elements (oxygen, sulfur) are mentioned briefly.

I. INTRODUCTION

Almost all information about isotope abundances in the interstellar medium (ISM) is obtained from observation of molecules. One must therefore attempt to ascertain what degree of confidence can be placed in molecular abundances to reflect the actual isotope ratios. Due to possible differences in the excitation and radiative transfer for the various isotopes, there also exists the problem of relating the observed quantity -- the intensity of the emission or absorption of radiation -- to the molecular abundance ratio. This has been treated extensively by others and will not be discussed here. For elements other than hydrogen, it has normally been assumed that the ratio of isotope abundances in a particular molecule is the same as that which

---

[*]Supported in part by the National Science Foundation of the U.S., Grant MPS 73-04781.
[**]Alfred P. Sloan Foundation Fellow.

Jean Audouze (ed.), CNO Isotopes in Astrophysics, 105-114. All Rights Reserved.

exists in the interstellar medium (ISM), e.g., $[^{13}C]/[^{12}]_{ISM} =$
$[^{13}CO]/[^{12}CO]$. In addition to providing information on isotope
abundances, a knowledge of isotope fractionation processes can also be
utilized to determine certain physical conditions of interstellar gas
clouds. As an illustration, upper limits to the electron density and
CO abundance in dense interstellar clouds is derived in the Appendix
from the DCO^+/HCO^+ ratio.

Although the session here is devoted to CNO isotopes, it is instructive for gaining appreciation of isotope fractionation effects in
interstellar molecules to consider the D/H ratio. For this case there
is little doubt that severe fractionation occurs.

II. HYDROGEN/DEUTERIUM FRACTIONATION

In _diffuse_ interstellar clouds ($A_v \lesssim 1$), the formation rate per
D-atom for HD exceeds the rate per H-atom for H_2 typically by a factor
$\sim 10^3$. Though destruction of H_2 and HD is due to photodissociation
through the same electronic transition, the rate is frequently greater
for HD by $\sim 10^4$. The net result is usually that $(HD/H_2) \lesssim (D/H)/10$ in
diffuse clouds. Thus the drastic differences in the molecular processes
for the two isotopes are not fully reflected in the molecular abundances
due to partial cancellation of an accidental nature.

In _dense_ interstellar clouds ($A_v \gtrsim 5$) which are utilized for most
isotope studies, the DCN/HCN and DCO^+/HCO^+ ratios are enhanced by
~ 200 (Penzias et al. 1976) and $\sim 10^4 - 10^5$ (Hollis et al. 1976),
respectively, if the actual (D/H) ratio is 2×10^{-5} as deduced in the
solar neighborhood. The galactic center seems so far to be unique in
having lower DCN and DCO^+ fractional abundances. This may either be
due to greater stellar processing of the ISM or differences in the
physical conditions that determine the chemical fractionation. In
addition, HDO (but not H_2O) has been detected. No other deuterated
forms have been found, though in most cases the sensitivity of these
null results is somewhat uncertain. The general type fractionation
processes which seem most likely to cause enhancement of deuterium
(e.g., Watson 1974, 1976) are similar to that which can alter the
$^{13}C/^{12}C$ ratio in ISM molecules. These are,

$$HD + XH^+ \rightarrow H_2 + XD^+ + \Delta E \tag{1}$$

where X is some molecule or radical. Another important possibility is
$YD^+ + X \rightarrow XD^+ + Y$, but its effect is qualitatively similar to equation (1).
It will not be discussed here. Typically, $\Delta E/k$ seems to be $\approx 300 - 500$ K
though exact values are not available (for $X = H_2$, a lower value is
suggested). In comparison, the kinetic temperature of the gas is
$\lesssim 50$ K. The product XD^+ in equation (1) can be observed directly, can
recombine with an electron to form a neutral, or can react to form
another type molecule. The enhancement of deuterium achieved in equation
(1) will normally be reflected in all of the products, as subsequent

reactions are unlikely to display isotope preferences. Reaction rates, ⟨cross section σ × velocity v⟩, have been measured for a number of reactions of the form (1) and are often rapid, $\langle \sigma v \rangle \approx 10^{-9}$ to $10^{-10} cm^3 s^{-1}$. Though it is unclear as yet which specific reaction(s) dominates in the dense interstellar clouds, a fractionation process of the type given in equation (1) will produce a (D/H) ratio in a particular molecule,

$$(D/H)_{mol} \approx \frac{g \langle \sigma v \rangle_x [HD]}{\langle \sigma v \rangle_x h[H_2] + \langle \sigma v \rangle_e [e] + \langle \sigma v \rangle_M [M]} \qquad (2)$$

Here the subscripts to the reaction rates ⟨σv⟩ indicate the type reaction with XH^+ -- "x", an isotope exchange as in equation (1); "e", dissociative recombination with an electron; "M", a chemical reaction with some neutral atom or molecule. Brackets [] indicate a number density. The probability that a deuterium atom in XD^+ of equation (1) will be passed along through the subsequent reaction chain to the product of interest is designated by g and $h = \exp(-\Delta E/kT)$. Likely choices for X that can cause the observed enhancement of deuterium in DCN and DCO^+ are CH_2 and H_2. Since these are not the same, the ⟨σv⟩ and ΔE appearing in equation (2) can be different. Thus the degree of enhancement of deuterium and its temperature dependence may not be the same in each molecule. This agrees with the observations. In the Appendix, equation (2) is utilized to determine information about conditions in the interstellar gas.

III. $^{13}C/^{12}C$ FRACTIONATION

(A) Proposed process

The alteration in properties which a molecule undergoes when carbon isotopes are exchanged is quite small in comparison with an exchange of hydrogen and deuterium. However, only "factor of two" fractionation effects of carbon isotopes are sufficient to confuse the interpretation whereas the observed effect for hydrogen isotopes is $\sim 10^2$ to 10^4. A mechanism by which carbon isotopes in molecules may become fractionated in relation to the true isotope ratio in the interstellar medium has been given in more detail by Watson et al. (1976) than will be presented here.

The basic idea is that a rate coefficient $\langle \sigma v \rangle = 2 \times 10^{-10} cm^3 s^{-1}$ has been measured for the reaction,

$$^{13}C^+ + ^{12}CO \rightleftarrows ^{12}C^+ + ^{13}CO + \Delta E \qquad (3)$$

($\Delta E/k = 35$ K) at laboratory temperatures. Rate coefficients for such ion-molecule reactions normally are constant in going to lower temperatures, so that the measured value is utilized for the forward reaction in equation (3) at interstellar gas temperatures ($\lesssim 50$ K). In order for reaction (3) to introduce a fractionation effect, competing reactions

must be comparable or slower. In <u>diffuse</u> interstellar clouds this
seems to be satisfied as the chief competing process in photodissociation
of the CO. For <u>dense</u> clouds, the main competitor is a chemical reaction
of the C^+ with another molecule. The abundance of such molecules is
uncertain since species which might reasonably be abundant, e.g., H_2O,
CH_4, and CO_2, are effectively unobservable at present in interstellar
clouds. Indirect evidence against high abundances for certain of these
species does exist, e.g., $[H_2O]/[CO] \lesssim 1/6$ (Snyder <u>et al</u>. 1977). Of
the observed molecules which have large rate coefficients for reaction
with C^+ (e.g., NH_3, HCN, H_2CO and OH) none have sufficiently high
abundances that these reactions will compete with equation (3). As a
result of equation (3) under conditions in <u>dense</u> clouds (i.e., ultraviolet radiation assumed negligible),

$$\frac{[^{13}C^+]}{[^{12}C^+]} \simeq \frac{[^{13}CO]}{[^{12}CO]} y \; , \tag{4}$$

with,

$$y = \frac{h + 10[M]/[^{12}CO]}{1 + 10[M]/[^{12}CO]} \; , \tag{5}$$

where [M] is the number density of molecules that have a rate coefficient
$\langle \sigma v \rangle_M$ for reaction with C^+ and $h = \exp[-35/T(K)]$.

     I will not go into the details of molecule formation schemes.
These are discussed, for example, by Dalgarno and Black (1976), Herbst
and Klemperer (1973) and Watson (1974, 1976). In most cases the exact
reactions that produce carbon-bearing molecules are uncertain. A
common feature of most proposals -- whether for <u>diffuse</u> or <u>dense</u> clouds,
gas phase or grain-surface processes -- is that the carbon which is in-
corporated into molecules exists more recently as C^+ than as CO. Thus
if no other isotope exchanges occur analogous to equation (3), the
^{13}C to ^{12}C ratio in most molecules other than CO will reflect that of
the ions $[^{13}C^+]/[^{12}C^+]$ given by equation (4). A few species may be
produced directly from CO ("CO-related"), and thus reflect the isotope
ratio for CO (HCO^+ and possibly H_2CO). Normally, in <u>dense</u> clouds, CO
is expected to contain most of the carbon in the gas. Thus the isotope
ratio in CO is the true value in the <u>gas</u>, and for most other molecules
the $^{13}C/^{12}C$ ratio is below this value according to equation (4) and the
above discussion. If there has been no significant increase in the
interstellar $^{13}C/^{12}C$ ratio since the formation of the solar system, if
most of the carbon in the gas is in CO, and if no other isotope selection
effects are important, then $[^{13}CO]/[^{12}CO] \approx 1/90$ and the "non-CO-related"
molecules which are made directly from C^+ will exhibit a lower value of
this ratio according to equation (4). Except possibly in one cloud
(toward Cas A, Matsakis <u>et al</u>. 1976), the observational evidence indicates a general enhancement of $^{13}C/^{12}C$ above the solar system value for
all molecules.

For conditions in <u>diffuse</u> clouds most carbon in the gas occurs as C^+, so that it is the $[^{13}C^+]/[^{12}C^+]$ ratio and the "non-CO-related" molecules such as CH^+ that reflect the true gas ratio. The ratio in CO will be enhanced under diffuse cloud conditions according to (Watson <u>et al.</u> 1976),

$$\frac{[^{13}CO]}{[^{12}CO]} = h^{-1} \frac{1+(\Gamma/2\times10^{-10}[^{12}C^+])}{1+(\Gamma/2\times10^{-10}h[^{12}C^+])} \frac{^{13}C^+}{^{12}C^+} \approx h^{-1} \frac{[^{13}C]}{[^{12}C]}_{ISM} \tag{6}$$

with the photodissociation rate $\Gamma(s^{-1})$ for CO and the C^+ density in cm^{-3}.

Before going on to discuss how $^{13}C/^{12}C$ may be enhanced in all molecules in <u>dense</u> clouds, note that other isotope exchanges analogous to equation (3) do seem to be unlikely (see Watson <u>et al.</u> 1976). The only exception is

$$^{13}C^+ + {}^{12}CS \rightleftarrows {}^{12}C^+ + {}^{13}CS + \Delta E, \tag{7}$$

though the rate coefficient for reaction has not been measured. Chemical reaction or charge-exchange would appear to dominate strongly over isotope exchange in other reactions that might be imagined with C^+.

Enhancement of the $^{13}C/^{12}C$ ratio in all gas molecules in <u>dense</u> clouds can occur because the vapor pressure of CO is higher than that of any other carbon-bearing molecule. Its vaporization temperature is $\sim 14\,°K$ at typical values of the partial pressure, whereas that of CH_4 is $\sim 20\,°K$. For other carbon-bearing molecules, the vaporization temperature is higher. Except very close to bright stars, temperatures for the dust-grains in interstellar clouds are $\sim 10\text{-}20\,°K$. There is strong, though perhaps not definitive, evidence from several sources in favor of depletion of heavy elements including carbon from the gas by freezing onto dust grains. In fact at the expected grain temperatures it is difficult to understand how depletion can be avoided. Due to the vapor pressure and observed abundance of CO, most of the carbon that freezes onto the grains is likely to hit the grain in a form other than CO. This form of carbon contains relatively more of the carbon-12 isotope, if it is also assumed that the abundance of "CO-related" molecules (excluding CO) is lower than "non-CO-related" ones. Thus relatively more carbon-12 is lost from the gas, so that the $^{13}C/^{12}C$ in the <u>gas</u> is <u>raised</u>. Re-cycling of carbon between CO and other molecules is normally thought to occur on a time scale that is shorter than the cloud lifetime. Then the $^{13}C/^{12}C$ ratio in all molecules can be increased above that which actually exists in the interstellar medium including dust-grains. Quantitative predictions for isotope ratio depend, in principle, on the history of the interstellar cloud. For the simplified case of (1) steady-state other than the depletion of carbon, (2) rapid recycling of carbon among the molecules of the gas in comparison with the rate of

depletion onto grains, and (3) a constant density of molecules of type
M, an analytic solution can be obtained,

$$\frac{[^{13}CO]}{[^{12}CO]} = \left[\frac{1}{f} \frac{1+(10f[M]/[^{12}CO])}{1+(10[M]/[^{12}CO])}\right]^{1-h} \left(\frac{^{13}C}{^{12}C}\right)_{ISM} \tag{8}$$

which along with equation (4) determines the $^{13}C/^{12}C$ for the two groups
of molecules ("CO-related" and "non-CO-related") in terms of a depletion
factor f for carbon. Here f = (carbon in the gas)/(carbon initially in
the gas). In Figure 1, the ratios obtained from equations (4) and (8)
are presented for representative values of the relevant parameters.
If instead of assumption (3) above, the ratio $[M]/[^{12}CO]$ remains
constant while depletion occurs,

$$\frac{[^{13}CO]}{[^{12}CO]} = (f)^{y-1} \left(\frac{^{13}C}{^{12}C}\right)_{ISM} \tag{9}$$

Isotope ratios deduced from equations (4) and (9) are presented in
Figure 2.

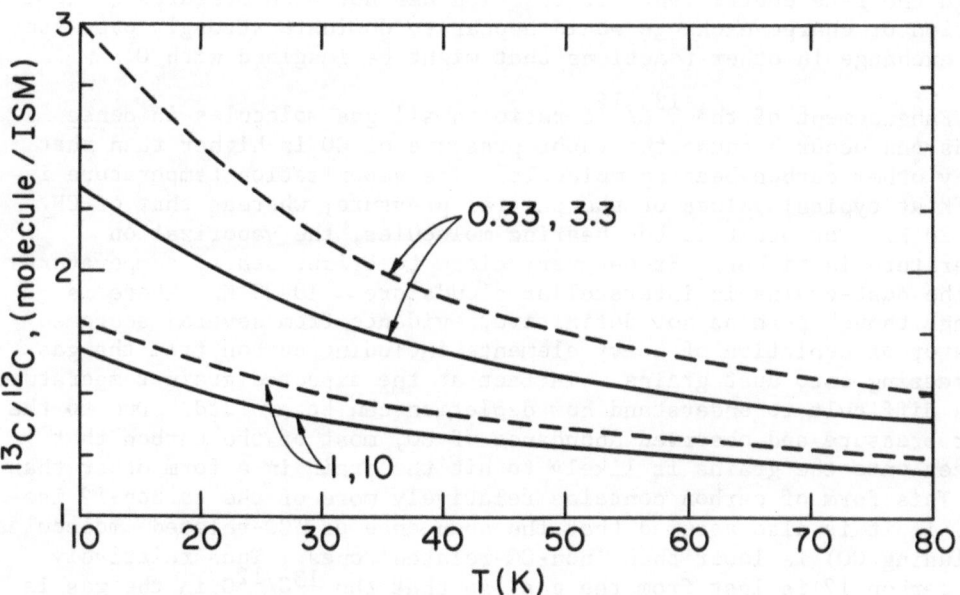

Figure 1. The $(^{13}C/^{12}C)$ ratio relative to the actual ISM value that is
predicted for the two groups of molecules -- CO-related including CO
(broken lines) and non-CO-related (solid lines) -- according to
equations (4) and (8). Curves are labeled by the parameters
$10f[M]/[^{12}CO]$, $10[M]/[^{12}CO]$.

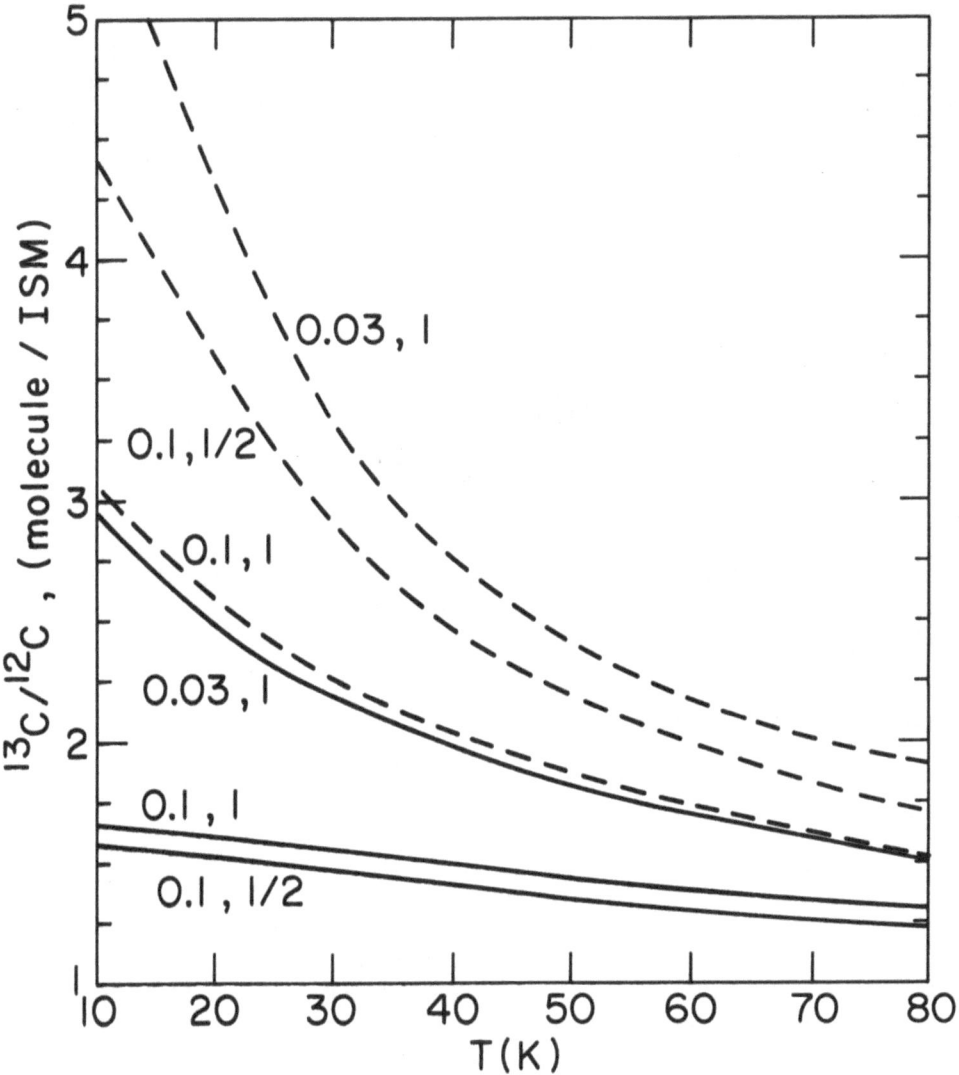

Figure 2. Same as Figure 1, except that fractionation is assumed to occur according to equations (4) and (9). Curves are labeled here by f, $10[M]/[^{12}CO]$.

The actual situation for interstellar clouds is probably more complex than represented by either of these simplifying approximations. Thus, a precise agreement between Figure 1 or Figure 2 and the observations should not necessarily be expected.

(B)   Uncertainties and observational tests

     The degree of fractionation that occurs in the dense clouds as a
result of the proposed mechanism is dependent upon several uncertainties
which include (1) the value of the rate coefficient for equation (3) at
temperatures $\gtrsim$ 50 K, (2) the abundance ratio $[M]/[CO]$, and (3) the
nature and amount of depletion of carbon.   Of these, the last is probably
the most serious.   For diffuse clouds, (1) seems to be the only reasonable uncertainty.

     A number of observational tests for fractionation will be possible
with refined observations.   (1) The "cleanest" test that fractionation
occurs to at least some degree is in the diffuse clouds, where the
fractionation process predicts that CO but not CH^+ will be enhanced
in carbon-13.   In dense clouds tests are less clear cut.   Variations in
physical conditions should produce variations in the $^{13}C/^{12}C$ ratio.
(2) Such variations can occur within a single cloud or (3) from cloud-
to-cloud.   (4) There should also be a difference in the isotope ratio
of $\gtrsim$ 25% between "CO-related" including CO and the "non-CO-related"
molecules (excluding possibly those with CS bonds).   An unambiguous
"non-CO-related" molecule in which ^{13}C can be detected is HCN.   Observational information relevant to the above tests has been presented by
several investigators at this symposium.

IV.   ISOTOPES OF OTHER ELEMENTS

     Exchange of oxygen isotopes analogous to equation (3),

$$^{18}O + C^{16}O \rightleftarrows {}^{16}O + C^{18}O + \Delta E \tag{10}$$

would lead to fractionation of oxygen isotopes in a manner analogous to
that in Section II for carbon if the rate for this reaction were large.
However, laboratory measurements (Jaffe and Kline 1966) find a quite
small rate coefficient at $\sim$ 300°K for reaction (10), as well as a substantial activation energy barrier.[*]   Three-body recombination data for
$CO + O + M \rightarrow CO_2 + M$ (Slanger et al. 1972) also support an activation energy
barrier for reaction (10).   This oxygen isotope-exchange can then be
ignored at interstellar cloud temperatures, as can reaction (10) with
O_2 or CO_2 substituted for the CO (see Jaffe and Kline 1966).   No other
exchange reaction seems likely to alter the oxygen isotope ratio.   For
sulfur isotopes the exchange,

$$^{34}S^+ + C^{32}S \rightleftarrows {}^{32}S^+ + C^{34}S + \Delta E \tag{11}$$

may be effective if it has a large cross section.

_____

[*]I thank W. B. DeMore for a helpful discussion about the likely rate
   coefficient for reaction (10).

APPENDIX

Most likely the strong enhancement in the DCO^+ abundance is due to,

$$H_3^+ + HD \rightleftarrows H_2D^+ + H_2 + \Delta E \tag{A1}$$

followed by,

$$H_2D^+ + CO \rightleftarrows DCO^+ + H_2 \tag{A2}$$

since the direct reaction,

$$HCO^+ + HD \rightleftarrows DCO^+ + H_2 + \Delta E \tag{A3}$$

apparently has a small rate coefficient.   Another possible contributor is a reaction between CH_2D^+ and oxygen,

$$CH_2D^+ + O \rightarrow DCO^+ + H_2 \tag{A4}$$

where enhancement of deuterium results from an exchange between CH_3^+ and HD analogous to equation (A1).   It seems less likely that reactions of the form (A4) dominate since this requires that HCO^+ not be produced by the mechanism (equation (A3)) postulated in the "ion-molecule" reaction scheme (Herbst and Klemperer 1973; Watson 1974; see also Snyder et al. 1977).

Irrespective of which reactions dominate for DCO^+, the resulting $[DCO^+]/[HCO^+]$ ratio is given by a relation of the form in equation (2). The DR21, NGC2264 and L134 ratio is ~ 0.1 to 1 for the observed antenna temperatures due to the $1 \rightarrow 0$ microwave transitions of these ions (Hollis et al. 1976).   Hollis et al. (1976) argue that at least for L134, the abundance ratio is approximately equal to the ratio of antenna temperatures (i.e., no appreciable line saturation).   From equation (2),

$$\langle \sigma v \rangle_e [e]/[H_2] + \langle \sigma v \rangle_M [M]/[H_2] < 4 \times 10^{-5} g \langle \sigma v \rangle_x / (D/H)_{mol} \tag{A5}$$

Then if the DCO^+ is due to equations (A1) and (A2) as seems most likely, the upper limits for L134 are $[e]/[H_2] \lesssim 4 \times 10^{-9}$ and $[M]/[H_2] \lesssim 2 \times 10^{-6}$. In this, the $T^{-\frac{1}{2}}$ dependence of $\langle \sigma v \rangle_e$ indicated (Leu et al. 1973) at laboratory conditions (down to 200 °K) has been utilized to extrapolate to ~ 20 K as suggested for L134.   This may be an overestimate.   Any saturation of the HCO^+ emission line in L134 would also raise the derived upper limits.

The low limit for $[e]/[H_2]$ is well below that derived by other methods, and has major importance for the gas/magnetic field coupling and the collapse of gas clouds in star formation.   Since CO reacts with H_3^+, the limit on M is equivalent to $[CO]/[H_2] \lesssim 2 \times 10^{-6}$. The CO abundance is of general interest, but also has special importance in studies of the mass of dense clouds and the mass distribution of material in the galaxy. These limits probably will be raised somewhat by the effects mentioned above. It is clear however that studies of DCO$^+$ are likely to provide valuable information on the abundances [CO] and [e] of interstellar clouds.

REFERENCES

Dalgarno, A. and Black, J. H.: 1976, Rep. Prog. Phys. 39, 573.

Herbst, E. and Klemperer, W.: 1973, Astrophys. J. 185, 505.

Hollis, J. M., Snyder, L. E., Lovas, F. J., and Buhl, D.: 1976, Astrophys. J. Letters, in press.

Jaffe, S. and Kline, S. F.: 1966, Trans. Faraday Soc. 62, 3135.

Leu, M. T., Biondi, M. A. and Johnsen, R.: 1973, Phys. Rev. A8, 413.

Matsakis, D. N., Chui, M. F., Goldsmith, P. F., and Townes, C. H.: 1976, Astrophys. J. Letters 206, L63.

Penzias, A. A., Wannier, P. G., Wilson, R. W., and Linke, R. A.: 1976, Astrophys. J., in press.

Slanger, T. G., Wood, B. J. and Black, G.: 1972, J. Chem. Phys. 57, 233.

Snyder, L. E., Watson, W. D., and Hollis, J. M.: 1977, Astrophys. J., in press.

Watson, W. D.: 1974, Astrophys. J. 188, 35.

Watson, W. D.: 1976, Rev. Mod. Phys., in press (Oct.).

Watson, W. D., Anicich, V. G., and Huntress, W. T.: 1976, Astrophys. J. Letters 205, L165.

IS THERE EVIDENCE FOR DEPLETION OF CNO IN THE INTERSTELLAR MEDIUM?

Gary Steigman
Yale University Observatory

In any discussion of interstellar abundances it is crucial to
know to what extent the gas phase abundances are representative of the
true interstellar abundances. Extensive interstellar absorption data
has been obtained in the last few years with the UV telescope on the
Copernicus satellite. It has been claimed that this data provides
evidence for depletion for many elements, including CNO (see the review
by Spitzer and Jenkins 1975 and references therein). My purpose here
is to point out that the conclusion that CNO are depleted from the
interstellar gas is based on an analysis of the data which is excessively
naive. I will conclude that there is no reliable evidence for depletion
of these elements in the relatively thin and diffuse clouds ($N_H \lesssim 10^{21} cm^{-2}$,
$n_H \lesssim 10^3 cm^{-3}$) studied in the Copernicus observations. It should be borne
in mind that I am not claiming that CNO are not depleted, rather, I
will simply argue there is no good evidence for their depletion. Although
attention is restricted here to CNO, my remarks will apply, in general,
to any abundance derived from saturated absorption features.

The basic problem in using interstellar absorption data to derive
abundances is that the desired quantity is a column density whereas the
observational quantity is an equivalent width. For unsaturated (or,
damped) absorption features many of the problems to be discussed below
do not arise, there is a reasonably unique relation between equivalent
width and column density and,therefore, a reliable abundance can be
derived. For features on the flat part of the curve of growth, in
contrast, there are several interrelated problems which complicate the
analysis. For such features the use of a simple curve-of-growth
analysis in passing from an observed equivalent width to a derived
column density is unreliable; the errors introduced by such an analysis
are unknown but likely to be large (anywhere from factors of a few to
several orders of magnitude).

For the sake of clarity, I have attempted to separate the difficulties associated with the derivation of column densities from equivalent
widths of saturated features. Keep in mind though, that the problems
outlined are not independent but, rather, like the other side of the

Jean Audouze (ed.), CNO Isotopes in Astrophysics, 115-118. All Rights Reserved.
Copyright © 1977 by D. Reidel Publishing Company, Dordrecht-Holland.

same coin.

    First, consider the idealized situation that, along a given
line-of-sight, there is only one absorbing "cloud" (for convenience,
I use "cloud" to refer to the absorbing region whether it is an HI or
an HII region and, whether it is a spatially compact region such as a
cloud or an extended region such as the intercloud medium). Further,
consider that the internal velocity distribution in this cloud is gaussian.
Even for this overly simplified case there can be large uncertainties
in deriving a column density from the equivalent width of a saturated
absorption feature by the single gaussian, curve-of-growth analysis. The
problem is that, for features on the flat part of the curve-of-growth,
a small uncertainty in the velocity dispersion corresponds to a large
uncertainty in the column density. The standard procedure has been to
use the velocity dispersion observed in an unsaturated feature of some
other element.   There are several problems associated with this procedure.
The relatively poor resolution ($\gtrsim 3 \mathrm{kms}^{-1}$) in the ultraviolet means that
clouds with low velocity dispersion (say, $\lesssim 1 \mathrm{kms}^{-1}$), separated in velocity
by only a few kms^{-1} will be unresolved. In such situations, the UV data
will provide a "single cloud velocity dispersion" which is too large,
leading to a derived column density which is too small. Even in the
case of one absorbing cloud there are problems. Inhomogeneities or
variations in ionization in the cloud may cause the velocity dispersion
determined from the absorption feature of some element to be different
from the velocity dispersion appropriate to the element under consideration. Even if the differences in velocity dispersion are small, the
corresponding differences in column density can be large. Such a
situation would resemble the superposition of several clouds at almost
the same velocity but with different internal velocity dispersions.
Still another complication is that the velocity dispersion is likely to
be partly thermal and partly turbulent. Since the thermal contribution
varies with the mass of the absorber, the velocity dispersion determined
from one element may not be appropriate to the absorber under consideration. Even though such differences probably will be small, I emphasize
again that a small uncertainty in the velocity dispersion translates into
a much larger uncertainty in the column density derived from features
on the flat part of the curve-of-growth.

    The assumption that the lines-of-sight to those stars studied in
the UV are dominated by a single absorbing region is unrealistic. Indeed,
it is known both observationally and theoretically that the single cloud
assumption must be wrong. From the high resolution optical absorption
studies of Hobbs and his co-workers (Hobbs 1969, 1971, 1974; Marschall
and Hobbs 1972) it is known that there are multiple HI components along
the lines-of-sight to most early type (O- and B-) stars. Further, there
must be circumstellar HII regions surrounding the O- and B-stars studied
in the UV (see, for example, Steigman, Strittmatter and Williams 1975;
Steigman 1977). Finally, there may be a distributed (in velocity as well
as spatially) contribution from an intercloud region (Hobbs 1976). It
is well known that the simple analysis breaks down in the presence of
multiple components (Nachman and Hobbs 1973). The failure of the simple

analysis has been explicitly exposed by Gomez-Gonzalez and Lequeux(1975) in a specific example chosen from the UV observations (see also Steigman 1977). The basic problem caused by the presence of multiple components can be conveniently understood as the combined effect of two interrelated facts. First, for a saturated feature (on the flat part of the curve-of-growth), a large increase in the column density corresponds to a small increase in the equivalent width. Hence, much material can be "hidden" along the line-of-sight leading to an underestimate of the true column density. Second, a region with high velocity dispersion and low column density can produce the same equivalent width as a low velocity dispersion, high column density region. Hence, high velocity dispersion regions will tend to dominate the observed equivalent width and much low velocity dispersion material can remain "hidden" along the line-of-sight.

Circumstellar HII regions have already been mentioned among the multiple components expected along the line-of-sight to a typical Copernicus star. For abundant elements such as CNO, in their abundant ionization states (e.g.:CII, CIII; NI, NII; OI) the contribution from circumstellar HII regions may dominate the observed equivalent widths (Steigman et al., 1975; Steigman 1977). The reason, quite simply, is the one alluded to above: HII regions have large velocity dispersions. For the stars studied in the UV, the HII region column density (N_{HII}) is typically ten percent of the total HI column density ($N_H \cong N_{HI} + N_{H_2}$). Although at first sight it might seem that these column densities should be uncorrelated, there is a strong selection effect at work. The larger N_H, the larger the reddening which requires that only very bright (and hot) stars be observed; such stars will have correspondingly larger HII regions in general. Thus, as the UV photons leave 0- and B-type stars, the first region they encounter is an efficient absorber. Such circumstellar HII regions may remove the appropriate photons so efficiently that the remaining low velocity dispersion HI material makes a negligible contribution to the observed equivalent width. Hence, circumstellar HII regions can effectively hide HI material along the line-of-sight. It should be noted that if, indeed, the CNO absorption features are dominated by such HII regions then, comparing the derived column densities with those of HI and H_2 is like comparing apples and oranges. Even if the derived column densities were reliable, the abundances determined this way would be meaningless.

The above remarks are quite general, referring to any saturated absorption feature on the flat part of the curve of growth. In particular, they are applicable to such abundant elements as CNO in their abundant ionization states. For typical values of the velocity dispersion and for solar abundances, individual HI clouds and HII regions are capable of producing saturated absorption features for the observed ions of CNO (CII, CIII; NI, NII; OI). The observed equivalent widths of these saturated features correspond to velocity intervals of several to several tens of kms^{-1}. Clearly, individual features are not being resolved and all the caveats discussed above apply. The column densities derived from such data are unreliable and so, therefore, are the abundances. The data provides no evidence for CNO depletion in the interstellar medium.

I would like to conclude with some general remarks about interstellar abundances, depletion and chemical evolution (meaning, in this case, the combined effects of nucleosynthesis and galactic evolution). It seems to me that the interpretation of interstellar absorption data has been biased towards finding depletions and against finding abundance variations. Of course, the two are intimately connected. Suppose, for example, that the column density of some element "X" is perfectly known (e.g.:$N(X)$ may be derived from unsaturated absorption features). Then, in terms of the depletion $\delta(X)$ (defined so that $\delta \leq 1$) and the true interstellar abundance $A(X)_{ISM}$, we have:

$$N(X) = \delta(X)A(X)_{ISM}N(H).\tag{1}$$

Comparing with the solar abundance of X we obtain:

$$\frac{N(X)/N(H)}{(X/H)_{\odot}} = \delta(X)A(X)_{ISM}/A(X)_{\odot}.\tag{2}$$

In all analyses of which I am aware, it is implicitly assumed that $A_{ISM} \equiv A_{\odot}$. [It is interesting that, until recently (see Dr. Watson's talk), the molecular absorption data has been interpreted in an entirely orthogonal manner. There, it is common to set $\delta \equiv 1$ and $A_{ISM} \neq A_{\odot}$].

It follows from (2) that depletions and abundance variations are tied together. Even if $N(X)/N(H)$ is less than $(X/H)_{\odot}$, it need not be the case that $A_{ISM} = A_{\odot}$. I strongly suspect that there has been an inhibition against finding (or, publishing) cases where $N(X)/N(H)$ exceeds the solar ratio. Rather than claim an "antidepletion", such data (if it exists) has probably been regarded as "unreliable".

REFERENCES

Gomez-Gonzalez, J. and Lequeux, J.: 1975, Astr. and Ap.,38,29.
Hobbs,L.M.: 1969, Ap.J., 157, 135.
_____ : 1971, Ap.J. (Letters), 170,L85.
_____ : 1974, Ap.J. (Letters), 188,L67.
_____ : 1976, Ap.J., 202,628.
Marschall, L.A. and Hobbs, L.M.: 1972, Ap.J.,173,43.
Nachman, P. and Hobbs, L.M.: 1973, Ap.J.,182,481.
Spitzer, L. Jr. and Jenkins, E.B.: 1975, Ann. Rev. Astr. and Ap.,13,133.
Steigman, G., Strittmatter, P.A. and Williams, R.E.: 1975, Ap.J.,198,575.
Steigman, G.: 1977, In Preparation ("On the Danger of Deriving Properties
     of the Interstellar Medium From Absorption Studies").

PART V

CNO ISOTOPE NUCLEOSYNTHESIS AND CHEMICAL EVOLUTION

THE CNO CYCLES*

GEORGEANNE R. CAUGHLAN
Montana State University, Bozeman, Montana, 59715, U.S.A.

Since the independent suggestions in 1938 of Bethe (1939) and von Weiszäcker (1938) that captures of protons by carbon and nitrogen isotopes were the source of energy generation in stars, the CNO cycles have developed into what might be considered by some a many-cycled monster rivalling the many-headed Hydra. However, those of us who are researchers in nucleosynthesis have found the reactions of the CNO nuclei a fertile field. I wish in this short paper to present the many reactions that may occur and to tell you what dependable rates we are able to make available.

The original CN cycle of Bethe and von Weiszäcker

$$C^{12}(p,\gamma)N^{13}(e^{+}\nu)C^{13}(p,\gamma)N^{14}(p,\gamma)O^{15}(e^{+}\nu)N^{15}(p,\alpha)C^{12}$$

supplies the main contribution of energy generation and synthesis of CN nuclei in main sequence stars more massive than the sun. However, nuclear research in the 1950's led Burbidge, Burbidge, Fowler, and Hoyle (1957) in their review paper that became the foundation for the study of nucleosynthesis, to the realization that in a very few proton captures by N^{15}, the NO cycle would occur with

$$N^{15}(p,\gamma)O^{16}(p,\gamma)F^{17}(e^{+}\nu)O^{17}(p,\alpha)N^{14}$$

producing the second cycle and making it necessary to study the CNO bi-cycle.

In the 1960's Caughlan and Fowler (1962, 1964, 1965) studied the data available on the nuclear reactions involved to obtain the rates of the reactions and the abundances of CNO nuclei that should be produced under equilibrium and approach to equilibrium conditions.

Since the early 1960's research on the rates of nuclear reactions

*Work supported in part by National Science Foundation Grant numbers (GP-9673 and AST 75-15854) at Montana State University.

and their effects on CNO abundances has demonstrated that hydrogen
burning by CNO nuclei is much more involved. Hoyle and Fowler (1965)
and Caughlan and Fowler (1965, 1967, 1972) showed that proton capture
by N^{13} must be considered under certain conditions leading to the
"fast" CN cycle

$$N^{13}(p,\gamma)O^{14}(e^{+}\nu)N^{14}.$$

This cycle also appears in the "hot" CNO processes considered by
Audouze and Fricke (1973), Audouze, Truran, and Zimmerman (1973),
Audouze (1973), Starrfield, Truran, Sparks, and Kutter (1972), and
Cowan and Rose (1975).

Original considerations of CNO cycles neglected the possibility
of radiative decay when O^{17} captures a proton because of the seemingly
greater probability of the $O^{17}(p,\alpha)N^{14}$ branch. The work of Rolfs and
Rodney (1974) which Rolfs (1976) described in the previous paper has
shown that the additional cycle

$$O^{17}(p,\gamma)F^{18}(e^{+}\nu)O^{18}(p,\alpha)N^{15}$$

must be included. This led to the CNO "tricycle" suggestions of Rolfs
and Rodney (1974, 1975) and Dearborn and Schramm (1974).

The equilibrium abundances reported in this paper were calculated
in a program using rates for the reactions published in Fowler, Caughlan,
and Zimmerman (1975) and included the five possible cycles shown in
Figure 1. Considering the complexity of the many possible cycles, it
would seem profitable to turn from the nomenclature of bi- or tri-cycle,
etc., to CNO cycles or CNO branches. At certain intermediate temperatures, the "fast" cycle is important when proton capture by N^{13} competes
with positron decay of N^{13}. If you wish, it is possible to identify
several bi-cycles in this diagram, and two additional branches must be
added to take into account the possible leak into F^{19} and Ne^{20} through
the $O^{18}(p,\gamma)F^{19}$ branch. As predicted by Rolfs (1974), the equilibrium
abundance data show that this leak has little effect on the CNO abundances because the $O^{18}(p,\alpha)N^{15}$ reaction is so much more probable than
the $O^{18}(p,\gamma)F^{19}$ reaction, and $F^{19}(p,\alpha)O^{16}$ is much more probable than
$F^{19}(p,\gamma)Ne^{20}$.

Although in the data presented here, the possibility of proton
capture by F^{17} and F^{18} was not included, I am presently working on
a program to take this into account using theoretically determined
rates for the $F^{17}(p,\gamma)Ne^{18}$ and $F^{18}(p,\alpha)O^{15}$ reactions that are included
in the multicycles shown in Figure 2. In the absence of experimental
data on those reactions, data available in Ajzenberg-Selove (1972) on
energy levels of Ne^{18} and Ne^{19} were used in the theory employed by
Fowler, Caughlan, and Zimmerman (1975) for similar situations.

Analysis of equilibrium abundances of CNO nuclei shows that when
the "fast" cycle is included, its effects are important in the region

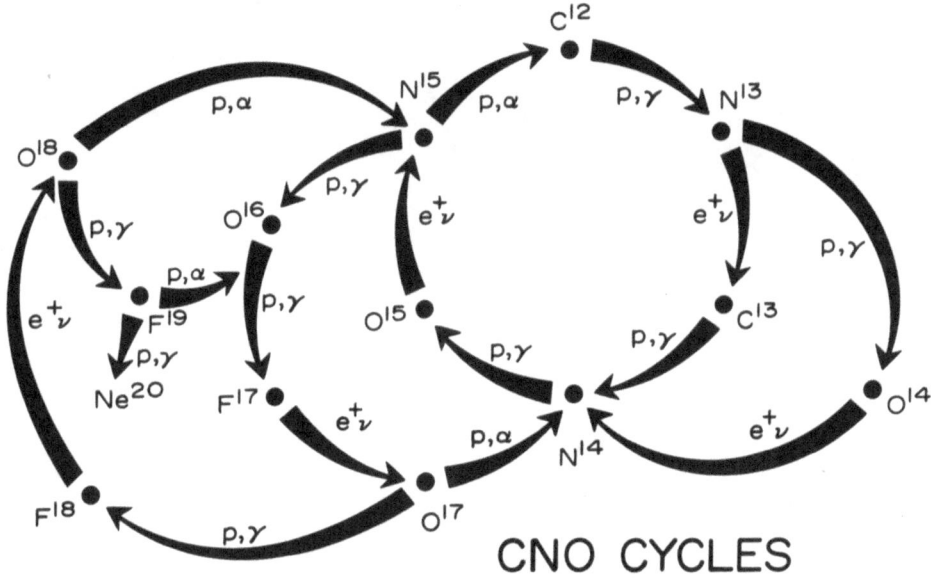

Fig. 1.   CNO Cycles

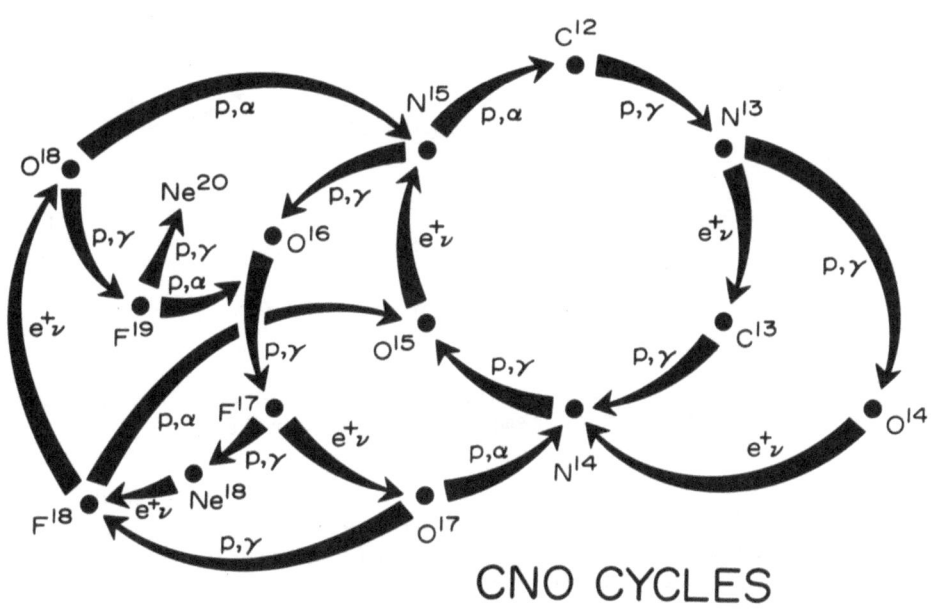

Fig. 2.   CNO Cycles including F^{17} and F^{18} proton captures.

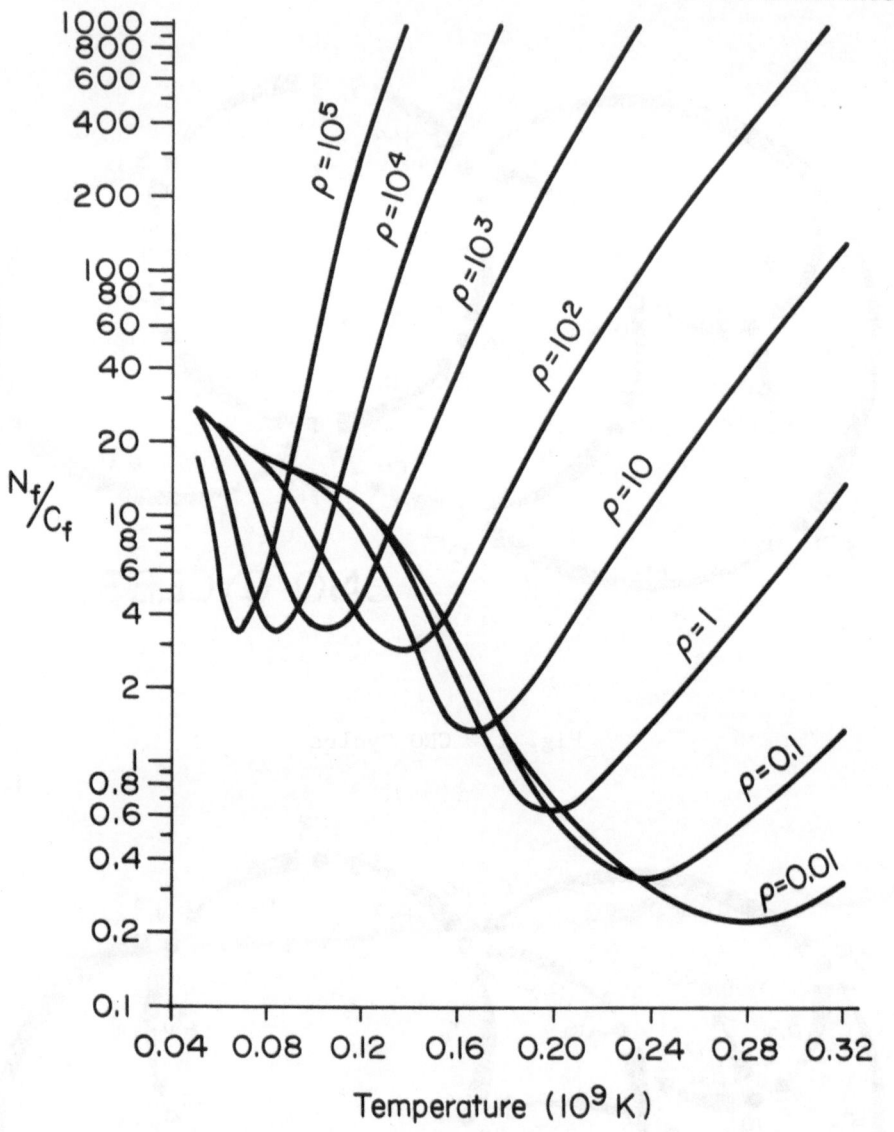

Fig. 3. Ratio of final nitrogen to final carbon abundances
as function of temperature and density.

T_9 = 0.06 to 0.24 depending on the density at the site of hydrogen
burning. Displayed in Figure 3 is the ratio of final nitrogen to
carbon (i.e. $(N_f^{14} + N_f^{15})$ to $(C^{12} + C_f^{13})$) after equilibrium abundances
of the unstable nuclei have decayed by positron emission. The ratio
is shown as a function of temperature and of density. It is evident
that at low temperatures, the nitrogen is more abundant than the
carbon since the "slow" cycle is in effect when positron decay of N^{13}
predominates. At intermediate temperatures, the competition between

proton capture and positron emission of N^{13} causes a minimum in the N_f/C_f ratio. At high temperatures the predominance of proton capture by N^{13} and the relatively long lifetimes of O^{14} and O^{15} for positron decay result in almost all the equilibrium abundances funneling into O^{14} and O^{15} which then produce N^{14} and N^{15} upon decaying after cessation of hydrogen burning. The ratio of N_f^{15}/N_f^{14} at high temperatures is 1.72 which is just the ratio of the mean lifetimes of O^{15} and O^{14}.

In Table IA ratios of the final abundances in "fast" CNO cycles at minimum N_f/C_f are shown. The calculations were based on a mass fraction of 0.745 for hydrogen. Also displayed is the ratio of the $N^{13}(p,\gamma)O^{14}$ rate to the $N^{13}(e^+\nu)C^{13}$ rate showing that they are comparable in the region of the minimum. The uncertainties of the rates of O^{17} and O^{18} reactions have little effect on the relative abundances of the CN nuclei to N^{14}. The high C_f^{13}/C^{12} and N_f^{15}/N_f^{14} ratios compared to the respective 0.29 and 4×10^{-5} values for C^{13}/C^{12} and N^{15}/N^{14} in the "slow" cycle are characteristic of the "fast" cycle.

The uncertainties of the O^{17} rates do affect the abundances of O^{17} and O^{18} as seen in Table IB. Those rates were determined by Rolfs and Rodney (1974) as Rolfs (1976) described in the previous paper. In the maximum rates the zero to one factor given in Fowler, Caughlan, and Zimmerman (1975) was chosen as one, and for the minimum rates for O^{17} that factor was taken as zero. The effect of the different rates is negligible for O^{16}/N_f^{14}, but, as one would expect, the O_f^{17}/N_f^{14} ratio is greater for the minimum O^{17} rates. At low densities in the neighborhood of N_f/C_f minimum, the maximum O^{17} rates have the greater O_f^{18}/N_f^{14} but at higher densities where the proton capture rates are increased in proportion to the density, the maximum O^{17} rates produce a smaller O_f^{18}/N_f^{14} than the minimum rates produce.

The relative abundances of the CN nuclei are shown in Table IIA for three characteristic temperatures, $T_9 = 0.120$, 0.140, and 0.300,

TABLE IA

RATIOS OF FINAL ABUNDANCES IN FAST CNO CYCLES AT N_f/C_f MINIMUM

ρX_H	T_9	N_f/C_f	C_f^{13}/C^{12}	C^{12}/N_f^{14}	C_f^{13}/N_f^{14}	N_f^{15}/N_f^{14}	$(p,\gamma)/(e^+\nu)*$
7.45E-03	0.278	0.231	0.807	3.35E+00	2.70	0.396	0.157
7.45E-02	0.237	0.345	1.36	2.10E+00	2.84	0.707	0.443
7.45E-01	0.198	0.640	2.26	8.23E-01	1.86	0.716	1.02
7.45E+00	0.166	1.36	3.46	2.60E-01	0.898	0.572	2.20
7.45E+01	0.136	2.86	4.68	8.35E-02	0.391	0.355	3.51
7.45E+02	0.103	3.50	5.55	5.17E-02	0.287	0.185	2.20
7.45E+03	0.083	3.45	7.01	4.32E-02	0.303	0.191	2.13
7.45E+04	0.068	3.38	8.91	3.58E-02	0.319	0.202	2.12

* $(p,\gamma)/(e^+\nu)$ is the ratio of the N^{13} rate to capture a proton to the N^{13} rate to decay by positron emission.

TABLE IB

RATIOS OF FINAL ABUNDANCES IN FAST CNO CYCLES AT N_f/C_f MINIMUM

ρ	ρX_H	O^{16}/N_f^{14}	O_f^{17}/N_f^{14}		O_f^{18}/N_f^{14}	
			a)	b)	a)	b)
0.01	7.45E-03	6.75E-02	4.25E-04	6.99E-04	1.92E-03	1.19E-04
0.10	7.45E-02	5.43E-02	5.74E-04	2.35E-03	8.00E-03	1.12E-03
1.00	7.45E-01	3.12E-02	4.65E-04	7.02E-03	1.19E-02	7.46E-03
10	7.45E+00	1.30E-02	3.06E-04	6.37E-03	1.03E-02	1.35E-02
10^2	7.45E+01	5.45E-03	1.88E-04	3.12E-03	6.28E-03	9.34E-03
10^3	7.45E+02	5.26E-03	2.23E-04	3.20E-03	2.62E-03	5.07E-03
10^4	7.45E+03	6.62E-03	2.97E-04	4.23E-03	7.80E-04	5.49E-03
10^5	7.45E+04	8.42E-03	2.23E-04	5.58E-03	1.86E-04	6.04E-03

a) Maximum O^{17} rates   b) Minimum O^{17} rates

and eight characteristic densities from ρ = 0.01 to 10^5 g cm^{-3}. Again, at higher densities large ratios of C_f^{13}/C^{12} and N_f^{15}/N_f^{14} appear showing the effect of proton capture competing with positron emission in N^{13}. Nitrogen exceeds carbon in all these because the temperatures are far enough removed from the regions in which N_f/C_f is minimum. The decrease in the C^{12} to N_f^{14} ratio with increasing density seen at T_9 = 0.140 is due to the increase in the (p,γ) rates

TABLE IIA

RATIOS OF FINAL ABUNDANCES IN FAST CNO CYCLES

ρX_H	N_f/C_f	C_f^{13}/C^{12}	C^{12}/N_f^{14}	C_f^{13}/N_f^{14}	N_f^{15}/N_f^{14}	TTAU sec
at T_9 = 0.120						
7.45E-01	10.9	0.358	6.75E-02	2.41E-02	9.89E-04	2.00E+05
at T_9 = 0.140						
7.45E-03	6.75	0.295	1.14E-01	3.38E-02	9.40E-05	3.27E+06
7.45E-02	6.64	0.318	1.14E-01	3.63E-02	6.54E-04	3.28E+05
7.45E-01	5.74	0.534	1.14E-01	6.11E-02	6.25E-03	3.37E+04
7.45E+00	3.11	2.02	1.13E-01	2.28E-01	6.15E-02	4.05E+03
7.45E+01	2.90	4.79	8.82E-02	4.22E-01	4.80E-01	7.37E+02
7.45E+02	14.2	5.65	2.52E-02	1.42E-01	1.37E+00	3.30E+02
7.45E+03	129	5.75	3.09E-03	1.78E-02	1.68E+00	2.87E+02
7.45E+04	1273	5.76	3.16E-04	1.82E-03	1.72E+00	2.82E+02
at T_9 = 0.300						
7.45E-01	7.62	3.72	7.63E-02	2.84E-01	1.74E+00	3.15E+02

and a consequent increase in N_f^{14}. Increasing the (p,γ) rates with density tends to increase the C_f^{13} to N_f^{14} ratio at low densities, but when the density is high enough (near $\rho = 100$ g cm^{-3} for $T_9 = 0.140$), the funneling to O^{14} leads to high final N^{14} abundances and a decreasing ratio of C_f^{13} to N_f^{14}. Also displayed in Table IIA is TTAU, the total cycle time in seconds to go through the full five cycles. Note that at high densities when the proton capture rates are very high and the abundances are essentially oscillating between O^{14} and O^{15}, the cycle time approaches 278 seconds which is just the sum of the 102 sec and 176 sec mean lifetimes of O^{14} and O^{15}.

The effects of the maximum and minimum O^{17} rates on the oxygen and fluorine plus neon abundances are displayed in Table IIB for the same characteristic temperatures and densities. The difference in the O^{17} rates has a negligible effect on the O^{16}/N_f^{14} ratio. As one would expect, the O_f^{17}/N_f^{14} ratio is less for the maximum O^{17} rates, but the effect decreases with increasing temperature and with increasing density as abundances funnel into O^{14}. Also, at higher temperatures, the nonresonant contribution to $O^{17}(p,\alpha)N^{14}$ becomes less effective than the resonant contribution to that rate. For maximum O^{17} rates, the O_f^{18}/N_f^{14} ratio increases with increasing temperature and with increasing density as more F^{18} is formed through $O^{17}(p,\gamma)F^{18}$, and its long mean lifetime of 9502 seconds allows much F^{18} to remain to form O^{18} on decay after hydrogen burning ceases. Examination of results of calculations from $T_9 = 0.05$ to 0.5 reveals that for the minimum O^{17} rates, O_f^{18}/N_f^{14} increases with temperature, but the increase ceases when $O_f^{18}/N_f^{14} = 7.52E-03$ at about $T_9 = 0.200$ for $\rho = 1$ g cm^{-3} when the $O^{17}(p,\alpha)N^{14}$ minimum rate exceeds that of $O^{17}(p,\gamma)F^{18}$ and a decrease in the O_f^{18}/N_f^{14} ratio occurs. For the maximum O^{17} rates, the (p,γ) and (p,α) rates are quite comparable in the temperature range where the minimum rates cause $O^{17}(p,\alpha)N^{14}$ to exceed $O^{17}(p,\gamma)F^{18}$. At temperatures above $T_9 = 0.220$, the $O^{17}(p,\alpha)N^{14}$ rate exceeds the $O^{17}(p,\gamma)F^{18}$ rate in both the maximum and minimum rates.

The leak of CNO abundances into fluorine and neon is not large as one can see from the ratio of $(FNe)_f/(CNO)_f$ from 10^{-9} to 10^{-14}. Also, that leak is not strongly affected by the differences in O^{17} rates.

Rolfs' calculations (1974) on the nonresonant contribution to the $O^{18}(p,\alpha)N^{15}$ reaction rate showed that the maximum value he obtained could be no more than a factor of about 10 high, hence, I chose the zero to one factor of Fowler, Caughlan, and Zimmerman (1975) as 0.1 for the minimum rate and the factor as one for the maximum rate of O^{18}. The differences in the O^{18} rates have a negligible effect on all abundances except O^{18}, F^{19}, and Ne^{20}.

The greatest effect of the different O^{17} rates is near the low-lying resonance at 0.066 MeV (LAB) which corresponds to a temperature of $T_9 = 0.047$. The contribution of that rate is uncertain, so it appears in the maximum O^{17} rate but not in the minimum rate. In Table III pertinent abundances are displayed for $T_9 = 0.045$ and 0.050

TABLE IIB

RATIOS OF FINAL ABUNDANCES IN FAST CNO CYCLES

ρX_H	O^{16}_f/N^{14}_f	O^{17}_f/N^{14}_f		O^{18}_f/N^{14}_f		$(FNe)_f/(CNO)_f$	
		a)	b)	a)	b)	a)	b)
at T_9 = 0.120							
7.45E-01	5.31E-03	8.93E-05	3.04E-03	1.65E-05	2.59E-05	1.77E-09	2.77E-09
at T_9 = 0.140							
7.45E-03	7.16E-03	7.41E-05	3.89E-03	1.10E-06	1.63E-06	1.70E-09	2.49E-09
7.45E-02	7.16E-03	7.43E-05	3.89E-03	1.10E-05	1.63E-05	1.69E-09	2.48E-09
7.45E-01	7.16E-03	7.63E-05	3.90E-03	1.10E-04	1.63E-04	1.65E-09	2.42E-09
7.45E+00	7.08E-03	9.52E-05	3.87E-03	1.09E-03	1.61E-03	1.37E-09	2.02E-09
7.45E+01	5.52E-03	2.29E-04	3.17E-03	8.52E-03	1.25E-02	7.54E-10	1.11E-09
7.45E+02	1.58E-03	5.07E-04	1.35E-03	2.44E-02	3.59E-02	1.69E-10	2.47E-10
7.45E+03	1.93E-04	6.02E-04	7.06E-04	2.98E-02	4.39E-02	1.94E-11	2.84E-11
7.45E+04	1.98E-05	6.14E-04	6.25E-04	3.05E-02	4.49E-02	1.97E-12	2.88E-12
at T_9 = 0.300							
7.45E-01	1.50E-03	3.34E-04	3.36E-04	4.60E-03	2.73E-04	3.42E-13	2.03E-14

a) Maximum O^{17} rates   b) Minimum O^{17} rates

TABLE III

RATIOS OF FINAL ABUNDANCES IN FAST CNO CYCLES

At $T_9 = 0.045$, $\rho X_H = 0.745$

N_f/C_f	C_f^{13}/C^{12}	N_f^{15}/N_f^{14}	O_f^{17}/N_f^{14}		TTAU
			a)	b)	sec
31.5	0.286	3.16E-05	1.20E-04	1.13E-02	1.52E+10

O_f^{18}/N_f^{14}								$(FNe)_f/(CNO)_f$			
a) c)	b) c)	a) d)	b) d)		a) c)	b) c)	a) d)		b) d)		
4.86E-11	4.58E-09	4.45E-10	4.19E-08		4.19E-11	3.91E-09	5.20E-11		4.85E-09		e
		7.56E-08	7.12E-06				1.96E-09		1.83E-07		f

At $T_9 = 0.50$, $\rho X_H = 0.745$

N_f/C_f	C_f^{13}/C^{12}	N_f^{15}/N_f^{14}	O_f^{17}/N_f^{14}		TTAU
			a)	b)	sec
27.8	0.286	3.03E-05	1.12E-04	9.52E-03	3.76E+09

O_f^{18}/N_f^{14}								$(FNe)_f/(CNO)_f$			
a) c)	b) c)	a) d)	b) d)		a) c)	b) c)	a) d)		b) d)		
6.09E-11	5.18E-09	4.39E-10	3.74E-08		3.78E-11	3.19E-09	4.60E-11		3.88E-09		e
		6.88E-08	5.68E-06				1.53E-09		1.29E-07		f

---

a) Maximum O^{17} rates   b) Minimum O^{17} rates
c) Maximum O^{18} rates   d) Minimum O^{18} rates
e) Uncertainty factor in $O^{18} = 0.10$
f) Uncertainty factor in $O^{18} = 0.00$

for a density of 1 g cm^{-3}. Ratios determined from maximum and minimum
O^{17} rates are marked in columns by a) and b) respectively. Those determined from maximum and minimum O^{18} rates are marked in columns by c)
and d) respectively. To see the greatest possible difference in O^{18}
rates, the ratios were also calculated with the zero to one factor in
O^{18} set equal to zero. The latter ratios are shown in lines marked
with an f. The characteristic high N/C and low C^{13}/C^{12} and N^{15}/N^{14}
ratios of the "slow" CN cycle can be seen. The minimum O^{17} rates produce a ratio of O_f^{17}/N_f^{14} that is about 100 times that produced by the

maximum O^{17} rates.   Comparison of O_f^{18}/N_f^{14} for the 0.1 minimum factor
shows a value at minimum O^{18} rates about 10 times that at maximum O^{18}
rates.   If the factor is taken as zero, the ratio of values at minimum
to those at maximum rates is more nearly a thousand.   However, lines
marked with an e probably have the more realistic results where the
factor was taken as 0.1.   The minimum O^{17} rates produce about 100 times
as much $F^{19} + Ne^{20}$ as the maximum O^{17} rates.   The different O^{18} rates
have little effect on $F^{19} + Ne^{20}$ if the realistic factor of 0.1 is
chosen for the minimum rate.

Analysis of the effects of the O^{17} and O^{18} rates at T_9 in excess
of 1 shows only a slight difference in $F^{19} + Ne^{20}$ and little difference in O_f^{18}/N_f^{14}.

To make a complete analysis of nucleosynthesis at the higher
temperatures, one must realize that reactions involving alpha-particle
captures must also be taken into account.   I am presently working on
such analysis to expand the work of Caughlan and Fowler (1964).   Several authors mentioned above have also looked at combined hydrogen and
helium burning.

I want to acknowledge the assistance of my daughter, Kerry C.
Travers, in the design of the pictures of the cycles in Figures 1 and 2
and to thank Professor William A. Fowler for many helpful discussions
during the preparation of this paper.

Work on this paper was supported in part by National Science
Foundation Grant numbers (GP-9673 and AST 75-15854) at Montana State
University.

"All our knowledge brings us nearer to our ignorance."

T. S. Eliot
The Rock, (1934)

REFERENCES

Ajzenberg-Selove, F.: 1972, Nucl. Phys. A190, 1.
Audouze, J.: 1973, in D. N. Schramm and W. D. Arnett (eds.), Explosive
   Nucleosynthesis, University of Texas Press, Austin, p. 47.
Audouze, J., Truran, J. W., and Zimmerman, B. A.: 1973, Astrophys. J.
   184, 493.
Audouze, J., and Fricke, K. J.: 1973, Astrophys. J. 186, 239.
Bethe, H. A.: 1939, Phys. Rev. 103, 434.
Burbidge, E. M., Burbidge, G. R., Fowler, W. A., and Hoyle, F.: 1957,
   Rev. Mod. Phys. 29, 547.
Caughlan, G. R., and Fowler, W. A.: 1962, Astrophys. J. 136, 453.
Caughlan, G. R., and Fowler, W. A.: 1964, Astrophys. J. 139, 1180.
Caughlan, G. R.: 1965, Astrophys. J. 141, 688.
Caughlan, G. R., and Fowler, W. A.: 1965, Astron. J. 70, 670.
Caughlan, G. R.: 1967, Am. J. Phys. 34, 69.

Caughlan, G. R., and Fowler, W. A.: 1972, Nature Phys. Sci. 238, 23.
Cowan, J. J., and Rose, W. K.: 1975, Astrophys. J. Letters 201, L45.
Dearborn, D., and Schramm, D. N.: 1974, Astrophys. J. Letters 194, L67.
Fowler, W. A., Caughlan, G. R., and Zimmerman, B. A.: 1975, Ann. Rev.
   Astron. and Astrophys. 13, 69.
Hoyle, F., and Fowler, W. A.: 1965, in I. Robinson, A. Schild, and
   E. L. Schücking, (eds.), Quasistellar Sources and Gravitational
   Collapse, University of Chicago Press, Chicago, p. 20.
Rolfs, C.: 1974, Private Communications.
Rolfs, C., and Rodney, W. W.: 1974, Astrophys. J. Letters 194, L63.
Rolfs, C., and Rodney, W. S.: 1975, Nucl. Phys. A250, 295.
Rolfs, C.: 1976, in J. Audouze (ed.), CNO Isotopes in Astrophysics,
   D. Reidel Publ. Co., Dordrecht-Holland.
Starrfield, S., Truran, J. W., Sparks, W. M., and Kutter, G.: 1972
   Astrophys. J. 176, 169.
von Weiszäcker, C. F.: 1938, Physik Z. 39, 633.

NEW CROSS SECTION MEASUREMENTS RELEVANT TO THE CNO NUCLEO-SYNTHESIS

C.Rolfs

Institut für Kernphysik,Münster,West-Germany

The reaction rates of nuclear processes in static hydrogen burning cannot be measured directly at the relevant stellar energies.In the usual procedure,the nuclear reactions are studied therefore over a wide range of beam energies and as low as possible in energy.These results are then extrapolated to stellar energies with the guidance of theoretical and other considerations.This procedure will be illustrated in the first example,the $^{15}N(p,\gamma)^{16}O$ reaction.However,the stellar reaction rates obtained through this extrapolation represent only lower limits if states exist in the compound nucleus near the proton threshold,which can alterate significantly the extrapolated reaction rates.The observed cross sections at higher beam energies are usually insensitive to the contribution of such threshold states.One has therefore to calculate the reaction rate of these threshold states, where all necessary parameters in the theoretical estimate (e.g. Breit-Wigner-formalism) have to be determined by experiments.Such a case will be described in the second example,the hydrogen burning of ^{17}O.

EXAMPLE 1: <u>Proton Capture by ^{15}N</u>

The cross section of the $^{15}N(p,\gamma)^{16}O$ reaction has been determined by Hagedorn[1] at E_p=800-1300 keV and by Hebbard[2] at E_p=220-700 keV.The results were analyzed[2] in

Jean Audouze (ed.), CNO Isotopes in Astrophysics, 133-144. All Rights Reserved.
Copyright © 1977 by D. Reidel Publishing Company, Dordrecht-Holland.

terms of two interferring $J^{\pi}=1^-$ resonances at $E_p=338$ and 1028 keV,with destructive interference in the energy region between the two resonances.From this analysis,an extrapolated astrophysical S-factor of $S(0)=26$ keV-b was reported[2].

In view of the reported large single-particle spectroscopic factors[3,4] for the ground state in ^{16}O,a direct capture ($\equiv$DC) component[5] of the type E1(s$\rightarrow$p) in the reaction $^{15}N(p,\gamma_0)^{16}O$ is expected,which can lead to additional interference effects of both $J^{\pi}=1^-$ resonances with the DC process.The DC process can result therefore in significant contributions to the capture process especially on the tails

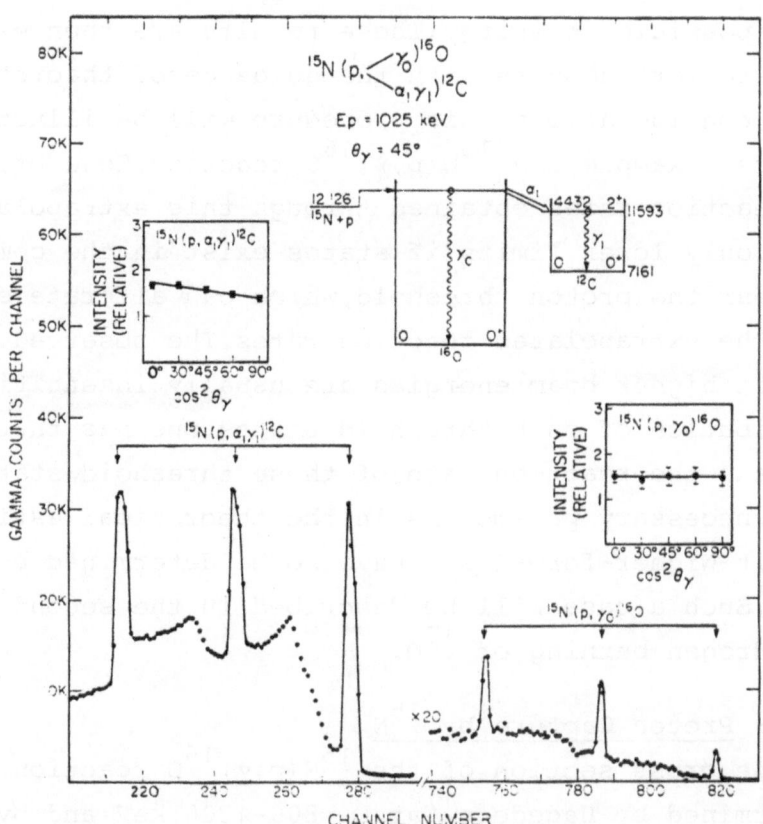

Fig.1.Relevant parts of the γ-ray spectrum obtained at $\theta_\gamma=$ 45° and $E_p=1025$keV.The insets show the γ-ray angular distributions of the 4432$\rightarrow$0(^{12}C) and R$\rightarrow$0(^{16}O) keV transitions observed at this beam energy.

of both resonances.In order to test this supposition,the $^{15}N(p,\gamma)^{16}O$ reaction was studied over a wide range of beam energies with the use of a Ge(Li) detector.The high energy resolution of a Ge(Li) detector also facilitates the search for other possible capture γ-rays to excited states in ^{16}O.

Proton beams of 10-70μA were supplied by the 0.6 and 2.6 MV accelerators at the Kellogg Radiation Laboratory.The TiN-targets (99% enriched in ^{15}N) were mounted in a target chamber,which allowed to set the Ge(Li) detector in close geometry.

Relevant parts of a sample γ-ray spectrum are shown in fig.1.It should be noted that below E_p=400keV,the $^{15}N(p,\gamma)^{16}O$ reaction is dominated by capture into the ground state of ^{16}O ($\geq$95%),in accordance with the previous assumption[2].The observed excitation functions of the $^{15}N(p,\alpha_1\gamma_1)^{12}C$ and $^{15}N(p,\gamma_0)^{16}O$ γ-ray transitions are shown in fig.2.Due to the isotropic or nearly isotropic angular distributions of both γ-ray transitions (insets in figs.1 and 2),small corrections had to be applied to the observed yields in order to obtain their total cross sections.

The sharp drop in the $^{15}N(p,\alpha_1\gamma_1)^{12}C$ yield curve below the 338keV resonance (fig.2) indicates,that -as in the case of the $^{15}N(p,\alpha_0)^{12}C$ reaction[2,6]- the two broad $J^{\pi}=1^-$ resonances at E_p=338 and 1028 keV interfere destructively below E_p= 300 keV.The extrapolation of these data to stellar energies gives -as an order of magnitude estimate- an S-factor of S(0)=0.1 keV-b.This contribution of hydrogen burning of ^{15}N is therefore negligible when compared with the reported value of S(0)=57 000 keV-b for the $^{15}N(p,\alpha_0)^{12}C$ reaction[2,6].

Fig.3 illustrates the observed data for $^{15}N(p,\gamma_0)^{16}O$ after conversion into the usual astrophysical S-factor.These data are dominated by the two $J^{\pi}=1^-$ resonances at E_p=338 and 1028 keV with a small contribution from the $J^{\pi}=1^+$ resonance at E_p=1640 keV.These data therefore were first analyzed in terms of the two $J^{\pi}=1^-$ resonances.If the usual energy-dependence of

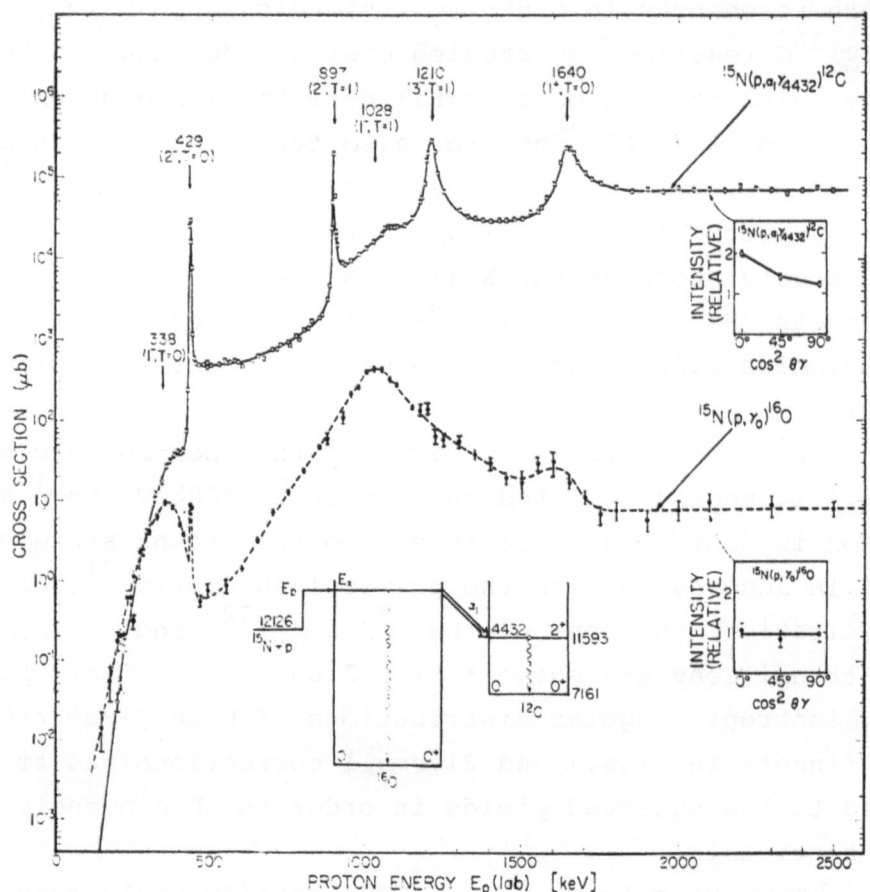

Fig.2.Total cross sections for the $^{15}N(p,\alpha_1\gamma_1)^{12}C$ and $^{15}N(p,\gamma_0)^{12}C$
   reactions are shown as a function of beam energy.The so-
   lid and dashed lines through the data points are placed
   to guide the eye.The insets show the γ-ray angular distributions observed at E_p=2.10 MeV.

the partial and total resonance widths as well as an interference term between the two resonances are taken into account,
an excitation function is obtained as indicated by the dashed
line in fig.3.The deduced S(0)-factor of 22 keV-b is in fair
agreement with the reported value[2] of 26 keV-b,obtained[2] from
a similar analysis.The calculated yield deviates however significantly from the data at the tails of the resonances,i.e.
the observed yield is a factor 2-3 higher than expected on the

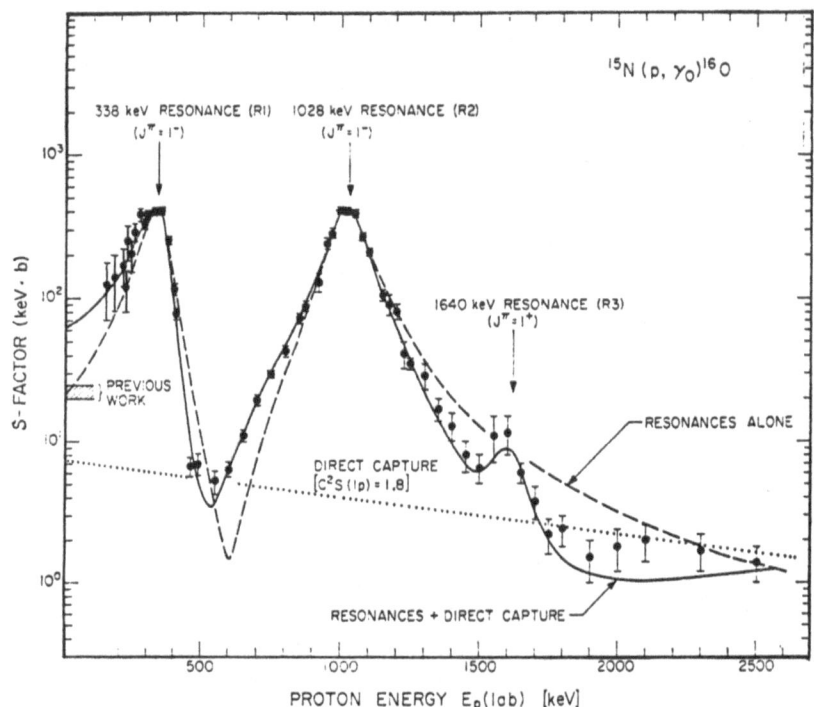

Fig.3.Astrophysical S-factor for the reaction $^{15}N(p,\gamma_0)^{16}O$.The
dashed line through the data points represents the re-
sults from an analysis solely in terms of the two $J^{\pi}=1^-$
resonances,revealing a S-factor of S(0)=22 keV-b.The
solid line represents the results from an analysis including in addition a direct capture component (dotted
line) in the capture process.This analysis reveals an
extrapolated S-factor of S(0)=64±6 keV-b.

low-energy side of R_1 and R_2 and a factor of 2-3 lower than
expected on the high-energy side of R_1 and R_2.These discrepan-
cies can however be removed,if a third component is included
in the capture mechanism,namely the expected DC→0 process
(solid line in fig.3).From this analysis,an astrophysical S-
factor of S(0)=64±6 keV-b is obtained,which is a factor of
2.5 higher than the previous value[2].This result implies that
every 880 cycles of the main CN cycle (rather than every 2200
cycles[2]) CN catalyst is lost through the $^{15}N(p,\gamma_0)^{16}O$ reactior

It is interesting to note that the S-factor at stellar energies is determined mainly from the features observed
at the tails of both resonances,especially in the beam energy
range E_p =400-900 keV,and not from the very low-energy data
points $(E_p \leq 250$ keV).An extrapolation of the low-energy data
points alone with their relative large errors would have allowed an S-factor in the range S(0)=10-100 keV-b.Since no
compound states in ^{16}O exist near the proton threshold,no
corrections have to be applied to the extrapolated reaction
rates from above.Further details of the proton capture by ^{15}N
are given in ref.7.

EXAMPLE 2: <u>Hydrogen Burning of ^{17}O</u>

The observed resonances in the $^{17}O(p,\alpha)^{14}N$ reaction
at $E_p \geq 500$ keV are too weak and too narrow to play a significant role at stellar energies.The low-energy tails of all these
resonances amount to an astrophysical S-factor of S(0)=1.0
keV-b.Since commonly (p,α) reaction rates on a given target
nucleus are much larger than associated (p,γ) rates,the contribution of the $^{17}O(p,\gamma)^{18}F$ capture reaction to the hydrogen burning of ^{17}O was completely neglected.Subsequent studies of this reaction confirmed,that the low-energy tails of
all observed resonances in $^{17}O(p,\gamma)^{18}F$ contribute very little
to the hydrogen burning of ^{17}O $(S_{p\gamma}(0)=0.01$ keV-b).Due to the
existence of compound states in ^{18}F near the proton threshold
(fig.4),Brown[8] suggested that the $^{17}O(p,\alpha)^{14}N$ reaction rate
is dominated in the stellar energy range by a resonance at
E_p =66 keV through the $E_x(J^\pi)$=5668(1$^-$)keV state.Interference
effects arising from the compound state at $E_x(J^\pi)$=5604(1$^-$)keV,
bound by 2 keV against proton decay (fig.4),also were taken
into account.For a calculation of this reaction rate,a two-
level resonance formula was assumed for the two J^π=1$^-$ states,
where each state was described by a Breit-Wigner shape.All
necessary parameters except the proton partial widths

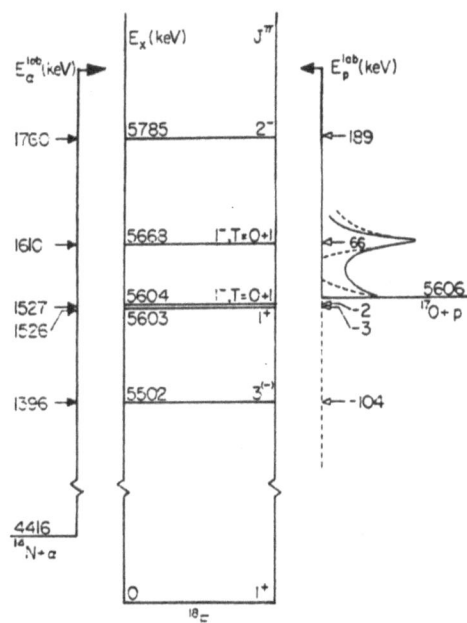

Fig.4.Compound states in ^{18}F relevant to stellar hydrogen
      burning of ^{17}O.Also shown is a schematic indication of
      the astrophysical S-factor both for constructive (solid
      line) and destructive (dashed line) interference bet-
      ween the two $J^{\pi}=1^{-}$ states.

(i.e.,proton-resonance energies,alpha and total widths) for
these calculations can be deduced from the $^{14}N(\alpha,\alpha)^{14}N$ and
$^{14}N(\alpha,\gamma)^{18}F$ data at the two respective resonances (fig.4).
from the systematics of the observed $^{17}O(p,\alpha)^{14}N$ resonance
strengths at $E_p \geq 500$ keV,Brown[8] assumed a reduced proton
width of $\theta_p^2(1=1)=0.007$ for the formation of both $J^{\pi}=1^{-}$ sta-
tes.With this value and the relation $\Gamma_p(E,1)=P_1(E)\theta_p^2(1)$,where
$P_1(E)$ represents the well-known transmission function,the
cross sections were then calculated for both constructive and
destructive interference between the two states.The calculated
rate for the assumption of destructive interference has since
then been used in many astrophysical calculations.
          In order to elucidate more this situation,an experi-
mental determination of the reduced proton widths $\theta_p^2(1)$ of

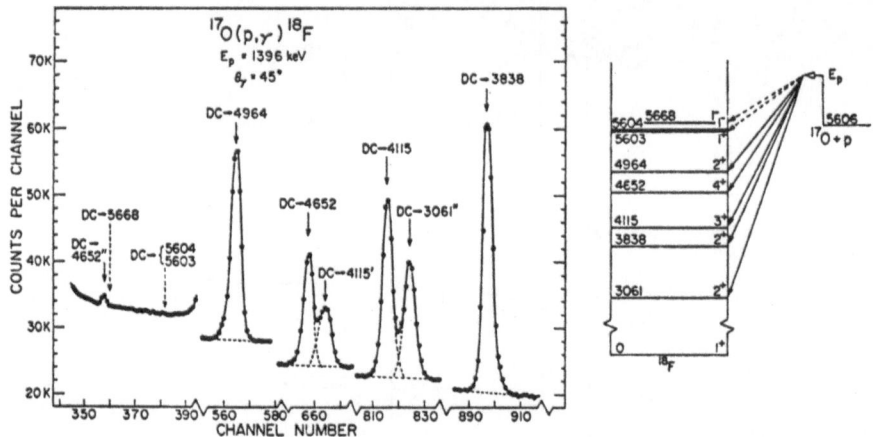

Fig.5.Relevant parts of the γ-ray spectrum for $^{17}O(p,\gamma)^{18}F$
      obtained with a Ge(Li) detector.

the 3 threshold states was desirable.This information can be
obtained[5] from the observed cross sections of the DC transitions to these states in the $^{17}O(p,\gamma)^{18}F$ reaction.A search
for these transitions was performed in the non-resonant energy region[5] of the above reaction (E_p=1.36-1.65MeV).The proton
beam of 70-160μA was supplied by the 2.6MV accelerator at the
Kellogg Radiation Laboratory.The WO_3-targets were enriched to
90% in ^{17}O.Fig.5 shows relevant parts of a γ-ray spectrum obtained at E_p=1.396 MeV.No DC transitions to the threshold states were observed in all the runs.The resulting upper limits
on their cross sections lead to reduced widths of $\theta_p^2(l=2)\leq0.002$,
$\theta_p^2(l=1)\leq0.0003$ and $\theta_p^2(l=1)\leq0.0001$ for the $5603(1^+)$,$5604(1^-)$ and
$5668(1^-)$ keV threshold states,respectively.With these values,
the cross sections and astrophysical S-factors have been cal-
culated.The results are shown in fig.6 for both constructive and
destructive interference between the two $J^\pi=1^-$ states.In the
case that both states are formed through f-wave capture rather
than p-wave capture,the present curves have to be scaled
down by a factor of 200.The influence of the $5603(1^+)$ keV state
on the stellar burning rates can be neglected (e.g. $S(E_p$=30
keV)≤ 0.03 keV-b).Since the γ-widths of the two $J^\pi=1^-$ states

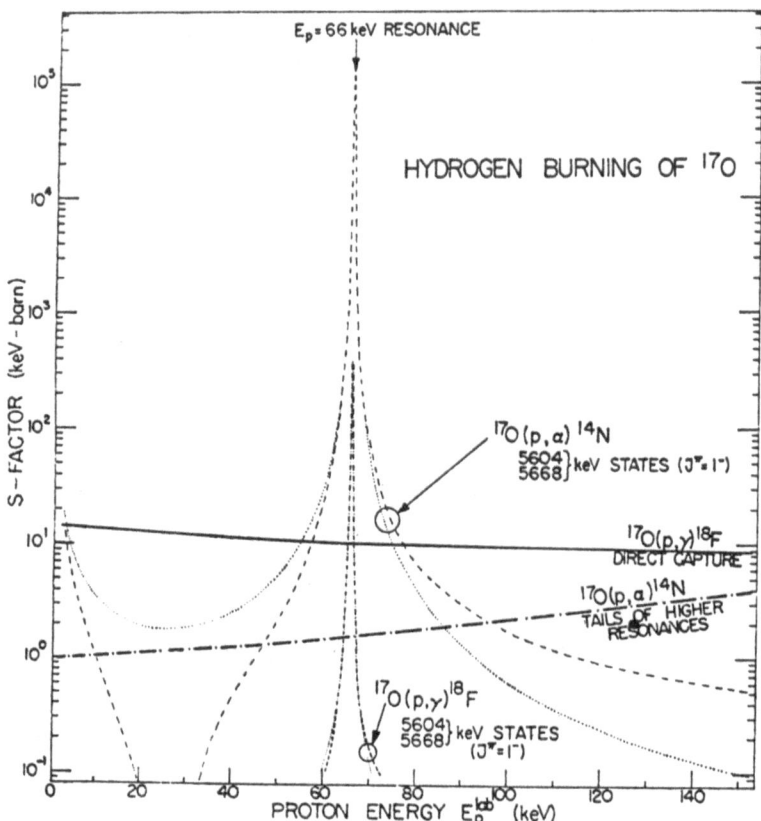

Fig.6.Astrophysical S-factor for the reactions $^{17}O(p,\alpha)^{14}N$
and $^{17}O(p,\gamma)^{18}F$.The S-factors for both reactions through
the $J^{\pi}=1^-$ states at $E_x=5604$ and 5668 keV are upper limits, where the dotted and dashed lines represent the
results for constructive and destructive interference
between the two states,respectively.Shown also are the
sum of the low-energy tails of all observed resonances
in $^{17}O(p,\alpha)^{14}N$ at $E_p\geq500$ keV (dash-dot line).The solid
line represents the S-factor for the DC process[5] in
$^{17}O(p,\gamma)^{18}F$ to final bound states.

---

are nearly two orders of magnitude smaller than their respective α-particle widths,the resulting S-curves of the
$^{17}O(p,\gamma)^{18}F$ reaction for these states are reduced by this
magnitude when compared to the $^{17}O(p,\alpha)^{14}N$ results (fig.6).

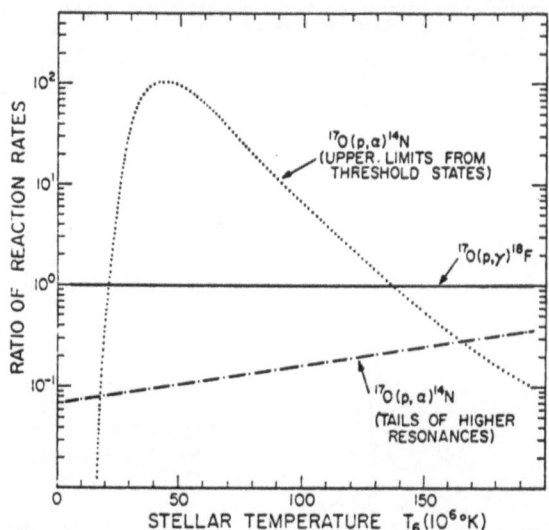

Fig.7.The reaction rates for $^{17}O(p,\alpha)^{14}N$ due to threshold states only (dotted line) and due to the low-energy tails
of the resonances at $E_p \geq 500$ keV only (dash-dot line)
are shown relative to the reaction rate for $^{17}O(p,\gamma)^{18}F$
(solid line) as a function of stellar temperature.

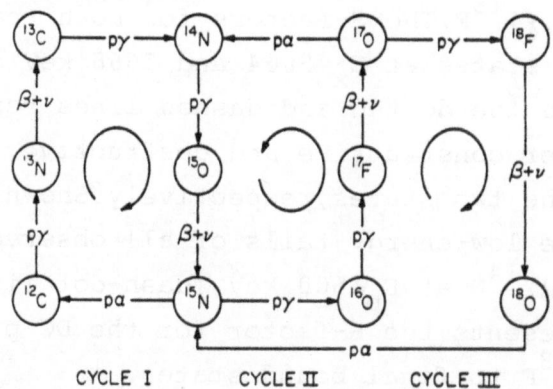

Fig.8.Illustration of the three cycles involved in the burning
of hydrogen via the CNO cycles.

As a result of these studies, the $^{17}O(p,\alpha)^{14}N$ reaction rate is reduced by a factor 60 when compared to previous estimates[8]. Away from the 66 keV resonance (fig.6), the total S-factor is influenced strongly by the direct capture process[5] in $^{17}O(p,\gamma)^{18}F$ to all the final bound states in ^{18}F. As a consequence, the open question of destructive or constructive interference between the two $J^{\pi}=1^-$ states is almost irrelevant, since at energies, where the two cases are most different, the reaction rate is dominated by the $^{17}O(p,\gamma)^{18}F$ process. If the S-factor curves of fig.6 are folded with a Maxwell-Boltzmann distribution, the stellar reaction rates are obtained. The resulting rate for the $^{17}O(p,\alpha)^{14}N$ reaction relative to that for $^{17}O(p,\gamma)^{18}F$ is shown in fig.7 as a function of stellar temperature. The low-energy tails of the observed resonances in $^{17}O(p,\alpha)^{14}N$ at $E_p \geq 500$ keV give rise to the lower curve (dash-dot line) and the upper limits in the reduced widths of the threshold states implie rates as indicated by the dotted line. The true rates will lie somewhere between these two curves. These calculations demonstrate that the $^{17}O(p,\gamma)^{18}F$ reaction cannot be neglected in the hydrogen burning of ^{17}O. If the subsequent hydrogen burning of ^{18}O (made after β^+ decay of ^{18}F) proceeds predominantly[9] through the $^{18}O(p,\alpha)^{15}N$ reaction, these studies lead to the conclusion that the CNO cycle is tricycling (fig.8). These results do not change significantly the energy release in the CNO cycle but affect principally the ^{17}O and ^{18}O elemental abundances.

REFERENCES

1) F.B.Hagedorn,Phys.Rev.108(1957)735

2) D.F.Hebbard,Nucl.Phys.15(1960)289

3) W.Bohne,H.Homeyer,H.Lettan,H.Morgenstern,J.Scheer and F. Sichelschmidt,Nucl.Phys.A128(1969)537

4) H.W.Fulbright,J.A.Robbins,M.Blann,D.G.Fleming and H.S.Plendl Phys.Rev.184(1969)1068

144   C. ROLFS

Continue

5) C.Rolfs,Nucl.Phys.A217(1973)29

6) A.Schardt,W.A.Fowler and C.C.Lauritsen,Phys.Rev.86(1952)527

7) C.Rolfs and W.S.Rodney,Nucl.Phys.A235(1974)450

8) R.E.Brown,Phys.Rev.125(1962)347

9) C.Rolfs and W.S.Rodney,Astr.J.Lett.194(1974)L63 and Nucl.
Phys.A250(1975)295

NUCLEOSYNTHESIS OF CNO ISOTOPES

James W. Truran
Department of Astronomy
University of Illinois

1. INTRODUCTION

The origin of the seven stable isotopes of carbon, nitrogen, and oxygen remains an unsolved problem. Current views as to the mechanisms of nucleosynthesis of these nuclei indicate a collective history which is extremely complex. Theoretical studies have revealed that quite varied astrophysical sites can contribute significantly to their production, including CNO-cycle hydrogen burning, core helium burning, thermal relaxation oscillations associated with double-shell-burning configurations in red giants, nova explosions, and supernova explosions. The relevance of detailed quantitative estimates of contributions from these diverse sites is clear from the discussions presented throughout this session: coupled to increasing observations of CNO elemental and isotopic abundances, such studies can yield important inferences concerning the history of the interstellar medium, the evolution of our galaxy and of other galaxies, the sources of cosmic rays, stellar evolution, and the mechanisms of nova and supernova explosions.

In this paper, the environments in which nucleosynthesis of CNO nuclei can occur are surveyed, and the results of recent numerical studies relating to these environments are summarized. A unifying discussion is presented in the concluding section.

2. ENVIRONMENTS FOR NUCLEOSYNTHESIS

2.1 Helium burning

Core helium burning most certainly represents the source of most of the ^{12}C and particularly ^{16}O present in galactic matter. The relative amounts of ^{12}C and ^{16}O formed remain uncertain due, primarily, to difficulties associated with the experimental determination of the $^{12}C(\alpha,\gamma)^{16}O$ reaction rate (Dyer, 1973; Dyer and Barnes, 1974; Fowler et al., 1975). Nevertheless, reasonable agreement with the solar system ratio $^{12}C/^{16}O = 0.55$ (Cameron, 1973) can be achieved. Arnett (1972)

Jean Audouze (ed.), CNO Isotopes in Astrophysics, 145-154. All Rights Reserved.
Copyright © 1977 by D. Reidel Publishing Company, Dordrecht-Holland.

has calculated the helium-burning evolution of massive stellar cores, utilizing a value $\theta_\alpha^2 = 0.085$ for the critical reduced width amplitude for the 7.12-MeV state in ^{16}O. Building upon these studies, Arnett and Schramm (1973) determined that the mass average of all material predicted to be ejected from stars with helium cores above $\sim 4 M_\odot$ (total masses $\gtrsim 10 M_\odot$), assuming a Salpeter initial mass function, gives a $^{12}C/^{16}O$ ratio consistent with the solar system value. They also found that the average star in this range produces carbon and oxygen in the ratio $^{12}C/^{16}O \sim 1$, consistent with the values determined for galactic cosmic rays (Shapiro and Silberberg, 1970).

While the origins of ^{12}C and ^{16}O, both in solar system matter and in cosmic rays, may be explained in this manner, it should be noted that carbon stars (Wallerstein, 1973; Scalo, 1974) can also contribute significant quantities of ^{12}C to the interstellar gas. Quantitative estimates of their contribution are difficult, due to uncertainties in stellar model calculations. In order to illustrate their possible importance, I will consider only a subclass of the carbon stars: those which also show substantial enhancements of s-process elements. I will further assume that the abundances observed in such stars are a consequence of thermal relaxation oscillations occurring in double-shell-source configurations (Schwarzschild and Harm, 1967). Calculations by Iben (1975a) of this evolutionary behavior in a 7 $M_\odot$ star provide the following intershell conditions: the mass fraction of ^{12}C is ~ 0.2, corresponding to a local enhancement of a factor ~ 60 relative to solar system matter; and the limiting s-process abundance level in this region represents an enhancement of a factor ~ 130 relative to solar (Iben, 1975b; Truran and Iben, 1977). It is clear that if this material, when mixed through the envelope following successive helium-shell flashes, provides the source of the s-process elements in red giants and ultimately in interstellar matter, then a substantial fraction of the ^{12}C may also be formed in such environments. The operation of the s-process mechanism of Ulrich (1973) yields a comparable prediction.

Helium burning episodes can also provide conditions appropriate for the formation of ^{18}O. Massive stars burning hydrogen by means of the CNO cycles convert initial abundances of carbon, nitrogen and oxygen to ^{14}N. During the helium burning phase, this ^{14}N is rapidly destroyed by the reaction $^{14}N(\alpha,\gamma)^{18}F$, following which ^{18}F will decay to ^{18}O. However, severe limitations on ^{18}O production are imposed by the fact that ^{18}O is itself destroyed at modest temperatures by the reactions $^{18}O(\alpha,\gamma)^{22}Ne$ and $^{18}O(\alpha,n)^{21}Ne$. The lifetime of ^{18}O against destruction by these reactions is shown as a function of temperature in Table I. Here, for purposes of illustration, I have assumed a density of 10^3 g/cc and a helium mass fraction $\chi_\alpha = 1$; the reaction rates are taken from Fowler, Caughlan, and Zimmerman (1975).

A comparison of these reaction lifetimes with those of core helium burning in massive stars suggests that little ^{18}O will survive. Helium burning in a core of 2 $M_\odot$ is initiated at a temperature of

TABLE 1
LIFETIMES FOR ^{18}O DESTRUCTION.

T	τ_{18_O}	T	τ_{18_O}
10^9 K	years	10^9 K	years
0.11	1.0 (8)	0.15	3.6 (4)
0.12	1.0 (7)	0.16	7.7 (3)
0.13	1.3 (6)	0.18	5.1 (2)
0.14	2.0 (5)	0.20	4.9 (1)

approximately 150 million degrees; more massive cores commence burning at even higher temperatures (Arnett, 1972). The helium-burning lifetime for the 2 $M_\odot$ core is $\sim$ 2 x 10^6 years, quite sufficient to assure that ^{18}O will be largely destroyed. Calculations of thermonuclear transformations associated with helium burning in massive stars by Couch and Arnett (1972) have confirmed this conclusion. Nucleosynthesis of ^{18}O in helium-burning cores is thus restricted to lower mass stars. It has yet to be demonstrated how any ^{18}O thus formed can find its way, unaltered, into the interstellar medium.

Comparable problems are encountered for shell helium-burning configurations. Helium-shell flashes associated with low mass ($\sim$ 0.5 $M_\odot$) carbon-oxygen cores result in peak temperatures which can allow ^{18}O to be formed and at least partially preserved in the inter-shell region. It is the further requirement that this intershell material be mixed to the surface regions of the star that presents difficulties. Such mixing is found to occur only after the core has grown substantially in size. Iben (1975a) finds that mixing results from the inward penetration of the surface convective zone for a core of 1 $M_\odot$; the same phenomenon occurs in the presence of more massive cores (Sugimoto and Nomoto, 1975). For these more massive cores, the increased flash temperatures are sufficient to assure ^{18}O destruction. Thus, the conditions for ^{18}O production in the shell and mixing of intershell matter to the surface are not simultaneous satisfied.

More promising environments for ^{18}O production are provided by explosive helium-burning episodes. Howard et al. (1971) have demonstrated that it can be formed as a result of the shock-induced thermonuclear processing of the helium zones in supernovae. These conditions predict comparable overproduction of ^{15}N (as well as ^{19}F and ^{21}Ne). ^{18}O production may also accompany helium core flashes in lower mass stars (Schwarzschild and Härm, 1966; Edwards 1970). Because serious questions concerning the dynamic behavior of these stars remain unresolved, detailed nucleosynthesis predictions are not possible. They nevertheless represent a possible low mass (non-supernova) source of ^{18}O, and may prove important to the interpretation of $^{16}O/^{18}O$ isotopic variations in the interstellar medium.

2.2 "Equilibrium" CNO-cycle hydrogen burning

The characteristics of CNO-cycle hydrogen burning in stars, for conditions under which steady-state or "equilibrium" burning configurations are realized, are well known (Caughlan and Fowler 1964; Caughlan 1965): (1) any initial concentrations of CNO nuclei present in the material will be converted almost entirely to ^{14}N; and (2) the $^{12}C/^{13}C$ ratio achieves an equilibrium value of approximately 3.4. Both of these features are direct consequences of nuclear rates. A recent nuclear experiment by Rolfs and Rodney (1974) revealed that the $^{17}O(p,\alpha)^{14}N$ reaction is much slower than previously assumed; this allows for some reduction of the $^{16}O/^{17}O$ ratio as well. Dearborn and Schramm (1974) determine a limiting ratio $^{16}O/^{17}O \gtrsim 200$ for steady-state conditions.

It is generally believed that such burning represents the source of most of the ^{14}N present in the galaxy. The helium zones in stars are expected to contain large concentrations of ^{14}N which has been formed in this manner. Significant mixing of ^{14}N through the entire hydrogen envelope occurs during red giant evolution in lower mass stars (Iben, 1964). Calculations of the chemical evolution of galaxies reveal that the abundance level of ^{14}N in our galaxy can be readily accounted for by these sources (Truran and Cameron, 1971; Talbot and Arnett, 1973; Audouze et al., 1975).

Concomitant production of ^{13}C and ^{17}O in concentrations consistent with solar abundances does not occur under equilibrium burning conditions. While the isotopic ratios are appreciably altered, the total elemental abundances of carbon and oxygen are far below that of ^{14}N. Nevertheless, the low $^{12}C/^{13}C$ and $^{16}O/^{17}O$ ratios observed in red giants are certainly to be attributed to CNO burning reactions: the $^{12}C/^{13}C$ ratio is found in some cases to approach the characteristic value $\sim$ 3.4. Incomplete CNO-cycle hydrogen burning can give rise to such ratios prior to the achievement of steady state burning configurations (Truran, 1972). Mixing of the envelopes of red giants can provide suitable environments for such burning. It is interesting to note that ratios $^{12}C/^{13}C \sim$ 4-10 in carbon and s-stars imply an enhancement of ^{13}C by a factor $\sim$ 30 with respect to solar system abundances, comparable to the average enhancement of the s-process elements themselves in s-stars. Assuming that these stars represent the source of the bulk of the s-process isotopes, it necessarily follows that a significant fraction of the ^{13}C also originates here. Red giant envelopes probably do not represent the dominant site for ^{17}O production: even the most favorable lower limit determined by Dearborn and Schramm (1974) for the $^{16}O/^{17}O$ ratio is hardly sufficient to assure that comparable enhancements of ^{17}O and ^{13}C can be formed.

2.3 High temperature and explosive CNO-cycle hydrogen burning

Hydrogen burning on CNO nuclei at temperatures above $10^{8°}$K can produce abundance distributions which differ significantly from those realized at lower temperatures. Hoyle and Fowler (1960) noted, for instance, the importance of the reaction $^{13}N(p,\gamma)^{14}O$ which competes favorably with $^{13}N(e^+\nu)^{13}C$ at such temperatures; Hoyle and Fowler (1965) have also shown that reactions like $^{14}O(\alpha,p)^{17}F$, $^{17}F(p,\gamma)^{18}Ne$, and $^{15}O(\alpha,\gamma)^{19}F$ can play an important role in nuclear transformations occurring at temperatures $T > 5 \times 10^8$ K. Quite generally, in contrast to burning at lower temperatures the rates of nuclear reactions induced by protons and alpha particles can compete favorably with position-decay rates under these conditions. The nuclear transformations which occur for temperatures in the range 10^8-10^9 K, for constant temperature and density conditions, have been discussed by Audouze et al. (1973) and will not be reviewed here.

High temperatures may be achieved in hydrogen-rich matter in several interesting astrophysical sites. Thermonuclear runaways in the accreted hydrogen-rich envelopes of white dwarfs are found to give rise to nova eruptions (Starrfield et al., 1976). Here, high temperature CNO burning results in significant overabundances of ^{13}C and, particularly, ^{15}N and ^{17}O in the nova ejecta (Starrfield et al., 1972), as their position-decay progenitors, ^{13}N, ^{15}O, and ^{17}F, are formed and preserved on the hydrodynamic expansion timescale. Supermassive stars (Appenzeller and Fricke, 1972; Fricke, 1973) similarly provide high temperature hydrogen-burning environments in which ^{13}C, ^{15}N, and ^{17}O may be formed (Audouze and Fricke, 1973). Shock processing of the envelopes of supernovae may also allow some production of these isotopes (see for example Howard et al., 1971), but the consequences of such burning episodes have not been thoroughly explored in the context of complete supernova envelope models.

A study of hydrogen-burning reactions proceeding at high temperatures under dynamic conditions has recently been completed (Lazareff et al. 1977). Illustrative calculations of such an "explosive" hydrogen-burning process are presented in Table II. Here, adopting the approach taken in previous explosive nucleosynthesis studies, one (1) specifies the initial temperature and density, (2) assumes an expansion timescale

$$\tau \sim \frac{446}{\rho^{\frac{1}{2}}} \text{ seconds,}$$

and (3) allows the matter to expand adiabatically. Such expansion profiles mimic reasonably well those characteristic of novae and supermassive stars. The initial composition was taken to be solar. As is evident from this table, ^{15}N and ^{17}O represent the most likely products of such burning stages; significant overabundances are realized over the range of temperature and density conditions explored.

3. DISCUSSION

The theoretical studies reviewed in the previous section support

TABLE 2

ENHANCEMENTS ACHIEVED IN "EXPLOSIVE HYDROGEN BURNING"

Ratios of Abundances to Solar-System Abundances

NUCLEUS	3×10^8 K		5×10^8 K		7.5×10^8 K	
	10^2 g/cc	10^4 g/cc	10^2 g/cc	10^4 g/cc	10^2 g/cc	10^4 g/cc
^{13}C	11.	.48	16.	2.6	11.	1.0
^{14}N	4.8	5.5	3.8	4.7	1.4	.33
^{15}N	920.	550.	1300.	2900.	920.	1200.
^{17}O	1400.	2300.	1100.	67.	1900.	130.
^{19}F				9.5		550.
^{21}Ne	24.	280.	4.7	210.	2.3	330.
^{22}Ne	11.	3.3	12.	5.2	4.5	42.
^{24}Mg	1.1	1.1	1.1	1.1	3.0	9.5

the following conclusions regarding the dominant modes of nucleosynthesis of the CNO isotopes:

^{12}C and ^{16}O represent the products of helium burning in the cores of massive stars which become supernovae.

^{14}N has its origin in equilibrium CNO-cycle hydrogen burning. It survives in unprocessed matter in the helium shells of massive stars ejected in supernova events. In lower mass stars, convective mixing during red-giant evolution enriches the entire envelope in ^{14}N (Iben, 1964), from whence it reaches the interstellar medium following less violent mechanisms of mass loss or planetary ejection.

^{13}C is produced in incomplete (non-equilibrium) hydrogen burning on ^{12}C in the envelopes of red-giant stars.

^{15}N and ^{17}O can be formed both in nova explosions and in supernova explosions. The relative contributions attributable to these sources are highly uncertain.

^{18}O is largely formed in the explosive ejection of the helium shells of supernovae.

The complexity of the thermonuclear origin of the CNO isotopes is apparent from these considerations. From the point of view of studies of galactic evolution and nucleosynthesis, this very complexity can be used to advantage. When coupled to observations of stellar and inter-

stellar medium abundances, quite stringent restrictions on the rates
of star formation and nucleosynthesis over the history of the galaxy
are imposed (Truran, 1973).

Two features of these diverse stellar environments are particularly
relevant to considerations of the chemical evolution of galaxies: the
masses and primordial metal abundances of the stars in which nucleosynthesis occurs. Both factors influence the time at which specified
nucleosynthesis products can first begin to contaminate the interstellar
gas. Contributions from low mass stars, for example, are substantially
delayed in their entry into the interstellar gas simply because their
hydrogen burning lifetimes are long. (Note that the average stars in
our galaxy--those of masses ~ 1 $M_\odot$ --have lifetimes $\gtrsim 10^{10}$ years and
thus cannot have influenced the primordial composition of the sun.)
Delayed entry into the interstellar medium of any nucleosynthesis
product dictates the need for a more rapid rate of growth to its solar
value. Variations in abundance ratios of the products of different
nucleosynthesis processes are thus to be expected.

Similar behavior results when such delay is necessitated by the
demand for primordial abundances of heavy nuclei to serve as seeds for
further nucleosynthesis. ^{14}N is a notable example of such a "secondary"
product of nucleosynthesis, requiring initial concentrations of carbon
and oxygen for its production in CNO-cycle hydrogen burning. An earlier
generation of massive stars must have enriched the interstellar gas in
^{12}C and ^{16}O prior to the formation of the stars in which ^{14}N was
produced. ^{12}C and ^{16}O, in contrast, represent primary nucleosynthesis
products: they can have been formed directly from hydrogen and helium
in the first generation of stars in our galaxy.

Given the primary character of ^{16}O and the secondary character
of ^{14}N, both high N/O ratios in metal-rich stars and N/O gradients in
galaxies can be understood. The effect is accentuated here by the
fact that increased production of ^{14}N relative to ^{16}O occurs at long
times (low rates of star formation) as the contributions from low mass
stars become dominant. Further predictions regarding CNO isotope ratios
are as follows:

$^{16}O/^{18}O$ ratios should show mild variations relative to
solar matter. ^{18}O represents a tertiary product, requiring
prior concentrations of ^{14}N for its production by $^{14}N(\alpha,\gamma)^{18}F(e^+\nu)$
^{18}O. It can be formed in a second stellar generation, however,
from ^{14}N produced in situ in the envelopes of massive stars
which evolve to supernovae. Because both ^{16}O and ^{18}O are formed
in massive stars, extreme $^{16}O/^{18}O$ ratios in the galactic center
are unlikely.

$^{12}C/^{13}C$ ratios are also expected to show mild variations.
The production of ^{13}C demands prior formation of ^{12}C; however,
^{13}C can still be formed in a first generation star when mixing
brings envelope hydrogen in contact with ^{12}C formed in situ in

the helium-burning shells of red giants. Low $^{12}C/^{13}C$ ratios observed in carbon (rich) stars testify to the operation of this mechanism. Contributions to ^{13}C from low mass stars (Dearborn et al., 1976) are limited by considerations of ^{14}N production in the same environments. Extremely low $^{12}C/^{13}C$ ratios in processed matter (the galactic center) are therefore not expected.

$^{14}N/^{15}N$ variations cannot be reliably predicted in view of uncertainties concerning the site of ^{15}N production. An extremely high value for this ratio in the galactic center would tend to support the view that supernovae, rather than novae, represent the dominant source of ^{15}N. I do not believe that a very strong statement can be made on the basis of current observations.

ACKNOWLEDGEMENTS

I wish to thank Icko Iben, Jr. for helpful discussions of several aspects of this review. This research was supported in part by the National Science Foundation, Grant AST73-05117.

REFERENCES

Appenzeller, I. and Fricke, K.: 1972, Astron. Astrophys. 21 , 285.

Arnett, W. D.: 1972, Astrophys. J. 176 , 681.

Arnett, W. D. and Schramm, D. N.: 1973, Astrophys. J. Letters 185 , L47.

Audouze, J. and Fricke, K.: 1973, Astrophys. J. 186 , 239.

Audouze, J., Lequeux, J., and Vigroux, L.: 1975, Astron. Astrophys. 43 , 71.

Audouze, J., Truran, J. W., and Zimmerman, B. A.: 1973, Astrophys. J. 184 , 493.

Cameron, A.G.W.: 1973, Space Sci. Rev. 15 , 121.

Caughlan, G. R.: 1965, Astrophys. J. 141 , 688.

Caughlan, G. R. and Fowler, W. A.: 1964, Astrophys. J. 139 , 1180.

Couch, R. G. and Arnett, W. D.: 1972, Astrophys. J. 178 , 771.

Dearborn, D.S.P., Eggleton, P. P., and Schramm, D. N.: 1976, Astrophys. J. 203 , 455.

Dearborn, D.S.P. and Schramm, D. D.: 1974, Astrophys. J. Letters 194 ,L67.

Dyer, P.: 1973, in D. N. Schramm and W. D. Arnett (eds.), Explosive
    Nucleosynthesis, Univ. of Texas Press, Austin, p. 195.

Dyer, P. and Barnes, C. A.: 1974, Nucl. Phys. A 233 , 495.

Edwards, A.: 1970, Monthly Notices Roy. Astron. Soc. 146 , 445.

Fowler, W. A., Caughlan, G. R., and Zimmerman, B. A.: 1975, Ann. Rev.
    Astron. Astrophys. 13 , 69.

Fricke, K.: 1973, Astrophys. J. 183 , 941.

Howard, W. M., Arnett, W. D., and Clayton, D. D.: 1971, Astrophys. J.
    165 , 495.

Hoyle, F. and Fowler, W. A.: 1960, Astrophys. J. 132 , 565.

Hoyle, F. and Fowler, W. A.: 1965, in I. Robinson, A. Schild, and E. L.
    Schucking (eds.), Quasi-Stellar Sources and Gravitational Collapse,
    Univ. of Chicago Press, Chicago, p. 17.

Iben, I., Jr.: 1964, Astrophys. J. 140 , 1631.

Iben, I., Jr.: 1975a, Astrophys. J. 196 , 525.

Iben, I., Jr.: 1975b, Astrophys. J. 196 , 549.

Lazareff, B., Audouze, J., Starrfield, S., and Truran, J. W.: 1977,
    "Hot CNO-Ne Cycle Hydrogen Burning II. Explosive Hydrogen Burning"
    (in preparation).

Rolfs, C. and Rodney, W. S.: 1974, Astrophys. J. Letters 194 , L63.

Scalo, J. M.: 1974, Astrophys. J. 194 , 361.

Schwarzschild, M. and Härm, R.: 1966, Astrophys. J. 145 , 496.

Schwarzschild, M. and Härm, R.: 1967, Astrophys. J. 150 , 961.

Shapiro, M. M. and Silberberg, R.: 1970, Ann. Rev. Nucl. Sci. 20 , 323.

Starrfield, S., Sparks, W. M., and Truran, J. W.: 1976, in P. P. Eggleton,
    S. Milton, and J. Whelan (eds.), "Evolution and Structure of Close
    Binaries", IAU Symposium No. 73, D. Reidel Publishing Company,
    Dordrecht, Holland (in press).

Starrfield, S., Truran, J. W., Sparks, W. M., and Kutter, G. S.: 1972,
    Astrophys. J. 176 , 169.

Sugimoto, D. and Nomoto, K.: 1975, Publ. Astron. Soc. Japan 27 , 197.

Talbot, R. J. and Arnett, W. D.: 1973, Astrophys. J. 186 , 69.

Truran, J. W.: 1972, in H. R. Johnson, J. P. Mutschlecner, and B. F.
     Peery, Jr. (eds.), Red Giant Stars, Indiana Univ. Press,
     Bloomington, p. 394.

Truran, J. W.: 1973, Comments Astrophys. Space Phys. 5 , 117.

Truran, J. W. and Cameron, A.G.W.: 1971, Astrophys. Space Sci. 14 , 179.

Truran, J. W. and Iben, I., Jr.: 1977, "On s-Process Nucleosynthesis in
     Thermally Pulsing Stars" (in preparation).

Ulrich, R. K.: 1973, in D. N. Schramm and W. D. Arnett (eds.), Explosive
     Nucleosynthesis, Univ. of Texas Press, Austin, p. 139.

Wallerstein, G.: 1973, Ann. Rev. Astron. Astrophys. 11 , 115.

GALACTIC EVOLUTION OF CNO ISOTOPES : THE CASE OF ^{15}N*

Jean AUDOUZE[1,2] ; James LEQUEUX[2] ; Brigitte ROCCA-VOLMERANGE[1] and Laurent VIGROUX[1,3].

[1]Laboratoire René Bernas du Centre de Spectrométrie Nucléaire et de Spectrométrie de Masse, 91406 ORSAY, France

[2]Département de Radioastronomie, Observatoire de Meudon, 92190 MEUDON, France

[3]DPhEP-ES, C.E.N. de Saclay, 91160 GIF-sur-YVETTE, France

I. INTRODUCTION

The study of the galactic evolution of the CNO isotopes using radioastronomical measurements of isotopic ratios in interstellar molecules is a fast-growing subject (see e.g. Wollman, 1973 ; Talbot and Arnett 1974, Audouze, Lequeux and Vigroux, 1975 = Paper I ; Vigroux, Audouze and Lequeux, 1976 = Paper II). Extensive discussions on this topic are presented in these papers. We will only summarize here the main conclusions that can be extracted from this work. Attention will be focused on the isotope ^{15}N for which very recent observations (Linke et al 1976) are available. This isotope has indeed very specific nucleosynthetic characteristics and the evolution models which can be derived from these observations are presented in this paper.

---

*Presented by J. LEQUEUX

II. A SUMMARY OF THE GALACTIC EVOLUTION OF THE CNO ISOTOPES

In this session,many observational results regarding different interstellar molecules have been presented (see e.g. the papers of Wannier, Encrenaz, Wilson in these proceedings). It appears in particular that the $^{12}C/^{13}C$ ratio observed in the interstellar medium is significantly lower (25-50) than in the solar system (89).

In spite of the possibilities of chemical fractionation discussed in particular by Watson (these proceedings), the fact that different molecules exhibit rather similar isotopic ratios, all far from the solar system ratio,has convinced us that there exists an actual difference between the interstellar and the solar system ratio. We interpret this difference by an enrichment of ^{13}C relative to ^{12}C by a factor of about 2.

The differences between the $^{12}C/^{13}C$ ratios observed in molecular clouds located at various galactocentric distances do not seem very significant although the $^{12}C/^{13}C$ ratio might be lower at the galactic center. Gradients in the isotopic or elemental ratios with galactocentric distance appear more important for the $^{15}N/^{14}N$ ratio as well as the O/H and N/H ratios : as we will see later the $^{15}N/^{14}N$ ratio decreases by a factor 5 when one goes from the solar neighborhood to the galactic center. In external spiral galaxies O/H and N/H ratios increase by factors respectively 5-40 when one goes from the outer to the inner regions (references are given in Paper I and II).

The interpretation of these differences have been made in Paper I and Paper II by taking into account the processing into stars (astration) of the interstellar matter in the frame of current models of chemical evolution of galaxies . It is assumed that the composition of the solar system represents that of the interstellar medium at the time of its formation i.e. 4.6×10^9 years ago. Therefore in this view the difference between the composition of the present interstellar medium and that of the solar system is a signature of the astration processes which have occured during the last 4.6×10^9 years. Similarly since the ratio gas / stars is much smaller in the galactic center ($< 10^{-3}$) than in the solar neighborhood ($\simeq 0.1$), the correlated difference in the rate of astration can be the cause of the observed gradients.

In this respect one can set a clear distinction between primary isotopes such as ^{12}C and ^{16}O and secondary isotopes such as ^{13}C and ^{14}N which are formed from preexisting primary isotopes. Primary isotopes appear first during the first stages of the galactic evolution but tend to level off later since they are the seeds of subsequent secondary nucleosynthesis. In contrast secondary elements appear later in the evolution but increase faster (as much as they are not destroyed subsequently themselves).

Therefore it is expected for example that the ^{12}C/^{13}C isotopic ratio decreases with time, at least when the amount of astration is not too large. This explains qualitatively why the ^{12}C/^{13}C ratio in the local interstellar medium is smaller than the ratio observed in the solar system. This also explains qualitatively the increase of ^{14}N/H and ^{16}O/H towards the center of spiral galaxies, and also why ^{14}N/H increases faster than ^{16}O/H.

In Paper I and II models have been designed to work out quantitatively the evolution of the ^{12}C, ^{13}C, ^{14}N and ^{16}O abundances. This has been made in the frame of the following assumptions :

a) The galaxy consisted at time t = 0 of pure gas (devoided of heavy elements, or already enriched in primary isotopes like ^{12}C and ^{16}C by a first generation of stars). The considered volume of the Galaxy is either isolated from outside, or continuously replenished with external gas (infall models).

b) The rate of star formation is proportional to the mass of gas σ (normalized to the total mass) present in the considered volume :

$(d\sigma/dt)_{\star} = -\nu\sigma$

The coefficient ν depends on the considered region but is taken independent of time. It is larger in the galactic center, a highly processed region, than in the solar neighborhood. This coefficient ν is actually adjusted to match the observed gas mass/total mass ratio (0.1 in the solar neighborhood and about 10^{-3} in the galactic center at the present time).

c) The initial mass function of stars (IMF) at their formation has been taken uniform in space and constant in time :

$$\psi(m) = \zeta \; (x-1) \; m^{-x} \quad \text{for } 1M_\odot < m < 60 \; M_\odot \text{ with } 1.30 < x < 1.75$$

and $\zeta = 0.25$ (ζ designates the mass fraction of the IMF corresponding to star $m > 1M_\odot$) ; the life time of a star has been taken proportional to the inverse cube of its mass (see paper II).

d) The mass and chemical composition in primary nucleosynthetic isotopes of the mass ejected by the stars at the end of their lives are similar to those given by Talbot and Arnett (1973).

e) ^{13}C is produced by the CNO cycle in red giants (see the contribution by D. Dearborn) ; we assumed $^{12}C_{initial}/^{13}C_{final} = 20$ for stars with $M < 5 \; M_\odot$, and 5 for stars of larger mass (actually, it turns out that the lower mass stars are the more efficient producers of ^{13}C during most of the galactic life). ^{14}N is also assumed to be produced by the red giants, but in much larger quantities and we have taken $^{12}C_{initial}/(^{14}N_{final} - ^{14}N_{final}) = 1.25$ for all stellar masses. ^{16}O and ^{12}C are assumed to evolve similarly.

f) In paper II, the simple but misleading assumption of instant recycling has been abandoned. Paper II shows how wrong this assumption is in particular when one consider strongly astrated regions such as the galactic center.

With these assumptions, we have been able to reproduce quantitatively the $^{12}C/^{13}C$ ratios in the solar system, the local interstellar matter and the galactic center, and also the N/H and the O/H gradients in spiral galaxies (which unfortunately are not yet determined precisely by current observations). Only models with a modest rate of infall of external matter (rate $< 5 \; 10^{-12} M_\odot$ of infalling gas per $M_\odot$ per year) give a good agreement between the calculated and the observed abundances. Models with larger infall rate ($\sim 10^{-11} M_\odot$ of infalling gas $M_\odot^{-1} yr^{-1}$ fail to reproduce the $^{12}C/^{13}C$ and N/H ratios in the galactic center (paper II).

The rest of this communication will be devoted the study of a rather interesting secondary element,^{15}N. Recently the group of the Bell Laboratories at Holmdel,N.J., U.S.A., has determined the abundance ratio of the rare isotopic species H^{12}C^{15}N and H^{13}C^{14}N in several molecular clouds including clouds at the galactic center (Linke et al, 1976). These measurements allow to study the production and the evolution of the ^{15}N isotope in the various regions of the Galaxy where it has been observed. As it will be seen in the next section, this isotope has indeed a rather particular nucleosynthetic history related to explosive hydrogen or helium burning.

III. THE NUCLEOSYNTHESIS OF ^{15}N.

As the other rare isotopes in the mass range $12 < A < 25,$ ^{15}N is a secondary product of nucleosynthesis. Its special interest comes mainly from the fact that, contrary to ^{13}C and ^{14}N for instance,it is not produced in appreciable amounts in the "cold" CNO cycles occuring at temperatures $< 10^8$ K in the inner zones of massive main sequence stars or in the hydrogen rich zone of the red giants : in fact the ^{15}N/^{14}N ratio which is reached in the cold CNO cycle equilibrium is at least 100 times lower than the solar system ratio (3.6×10^{-3}). In the cold CNO cycle ^{15}N is destroyed by the reaction ^{15}N (p, α) ^{12}C which is the fastest of the cycle.

The isotope ^{15}N is easily produced in the so-called "hot" CNO cycle occuring at T > 10^8 K where the slowest reaction is ^{15}O (β^+) ^{15}N (see eg. Audouze et al , 1973). It can be also produced in the explosive nucleosynthesis of helium-rich zones (see eg. Howard et al, 1971). Likely astrophysical sites for the hot explosive CNO cycle may be the novae (see eg. Starrfield et al, 1972 and 1974). For these authors this cycle of reactions actually triggers the nova outburst. Their prenova models are such that T $\gtrsim$ 1.4 10^8K, $\rho \sim 10^3$ g cm^{-3} and the initial material is enriched into CNO elements (this last hypothesis is being questioned by Prianilk et al, 1976). In favour of these models the observations of the ^{15}N/^{14}N ratio in Nova Herculis 1934 made by Sneden and Lambert (1975)

compared to the solar system value seems to show that ^{15}N is
enriched relative to ^{14}N. Explosive hydrogen and helium burning might
also occur in supernovae if temperature-density conditions $T > 10^8$K, $\rho >$
0.1 g cm^{-3} are achieved. Similarly if supermassive objects $(M \simeq 10^5 M_\odot)$
exist, Audouze and Fricke (1973) have shown that the hot CNO cycle can
trigger their explosion. In any circumstance the nucleosynthesis of
^{15}N seems to be related to sites where explosive hydrogen and/or helium
burning can take place, the most likely being the novae.

IV. THE OBSERVATIONS REGARDING THE INTERSTELLAR ^{15}N

A first series of measurements of the $H^{12}C^{15}N/H^{13}C^{14}N$ ratio has
been made for Orion A by Wilson et al, (1972) ; Clark et al, (1974) ;
and Wannier et al, (1975), who found respectively values of 0.38 ± 0.12, 0.22
and 0.28   which are comparable to the solar system value of 0.33.

Very recently Linke et al, (1976) found values of $H^{12}C^{15}N/H^{13}C^{14}N$
for four molecular clouds rather close to the Sun : NGC 7538 (0.13),
Ori A (0.28), NGC 6334 (0.12) and W51 (0.20). The large variations from
source to source are about 30%, larger than the errors in the observations
and the uncertainties in the interpretation of the line intensity. Linke
et al, (1976) think that they are due to relative abundance variations in
the nitrogen isotopes. However, there is no doubt that the average local
interstellar $^{12}C^{15}N/^{13}C^{14}N$ ratio : 0.19, is smaller than the solar system
value : 0.33.

In the two molecular clouds associated to the galactic center
(Sgr B2 and Sgr A),   observations provide an upper limit
$^{12}C^{15}N/^{13}C^{14}N < 0.026$ (1σ) very much smaller than both the solar system
and the local interstellar ratios. Since the $^{12}C/^{13}C$ ratio is known not
to change by factors larger than $\simeq 2$ between the local and the galactic
center interstellar gas (Wannier et al, 1975), this strong variation is
indeed due to a strong decrease by factors > 3-4 of the $^{15}N/^{14}N$ ratio
when one goes from the solar neigborhood to the galactic center. The purpose of this note is then to offer some explanations of this strong variation in terms of chemical evolution models.

V. THE EVOLUTION OF THE ^{15}N/^{14}N RATIO

To derive from the above ratio the behaviour of ^{15}N it is necessary to have independant information on the ^{12}C/^{13}C ratio and on the ^{14}N abundance. The ^{12}C/^{13}C ratios and the ^{14}N abundances have been estimated from simple galactic evolution models by Vigroux et al,(1976) and shown to be in agreement with the observed values of ^{12}C/^{13}C in our Galaxy and of ^{14}N/H in external galaxies. It is important to remember that in these models ^{13}C and ^{14}N are assumed to be fully secondary elements synthetized in the hydrogen-rich zones of red giants of any mass. Table 1 gives the relevant quantities coming from a model which gives a good agreement with observations, and the derived values for the ^{15}N abundance in the local and the galactic center interstellar gas (the evolutionary models considered here do not include the effects of infall of external gas ; such effects will be analyzed later). This table confirms the suggestion made by Linke et al, (1976) that the strong reduction of the ^{12}C^{15}N/^{13}C^{14}N ratio in the galactic center is mainly due to a strong enrichment in ^{14}N and not to a strong depletion of ^{15}N.

For the purpose of this still crude analysis the evolution models of Paper II have been adapted to give some estimate of the ^{15}N abundance evolution.

As it has been seen in section III, if the nuclear processes responsible for the ^{15}N production are reasonably understood, the same cannot be said of the nature and evolution of the progenitor objects. We shall then calculate the evolution of ^{15}N by making crude assumptions on the evolutionary properties of these objects. The method is described in Vigroux et al, (1976) and the assumptions are the same ; we used for the solar neighborhood an astration parameter $\nu = 0.2$ per 10^9 years (ν is defined by the equation $(d\sigma/dt)_* = - \nu\sigma$ relating the rate of star formation $(d\sigma/dt)_*$ to the gas mass σ); for the galactic center we take $\nu = 2$ per 10^9 years. We assume that the solar system ^{15}N/^{14}N ratio is representative of the ratio in the interstellar medium 4.6×10^9 years ago, and we normalize the rate of ^{15}N production so that its solar abundance is reproduced by the calculations. We choose 13×10^9 years as the age of the Galaxy. In most cases we have calculated the evolution of ^{15}N by assuming

TABLE I

Observed $^{12}C^{15}N/^{13}C^{14}N$ ratios, computed ^{12}C, ^{13}C, ^{14}N mass abundances ("best" model of Paper II) and $^{15}N/H$ mass abundances inferred to match the previous computations. It can be noted that ^{15}N might have increased in the galactic center while the $^{12}C^{15}N/^{13}C^{14}N$ ratio has significantly decreased. This is due to the strong ^{14}N gradient obtained in our previous calculations to match the observed spatial gradient in N.

	$^{12}C^{15}N/^{13}C^{14}N$ observed	$^{12}C/^{13}C$ calculated (Paper II)	$^{12}C/H$ 10^{-3}	$^{13}C/H$ 10^{-5}	$^{14}N/H$ 10^{-3} per mass (PaperII)	$^{15}N/H$ Inferred, per mass 10^{-6}
Solar system	0.33	89	4.4	5.0	1.1	4
Interstellar matter in solar neighborhood	0.19 (average)	57	6.2	10.0	2.8	9
Interstellar matter in galactic center	< 0.026 (1σ)	19	3.6	18.	11.	<15

that either it is destroyed by stellar processing after its production, or it is not destroyed. The results obtained with these two extreme assumptions are not very different ; here we present only those obtained assuming destruction of ^{15}N by stellar processing.

A - $\underline{^{15}\text{N is produced at a rate uniform in space per unit total mass, and}}$
$\underline{\text{independent of time}}$.

With this hypothesis, the calculated $^{12}\text{C}^{15}\text{N}/^{13}\text{C}^{14}\text{N}$ ratio in the galactic center is at least equal to the solar system value and therefore at least 10 times larger than the observed upper limit. The agreement is not better if we assume that the ^{15}N production starts at various times after the birth of the Galaxy. Thus this model of ^{15}N production is clearly not compatible with observations.

B - $\underline{\text{The rate of }^{15}\text{N production is proportional to the rate of mass ejection}}$
$\underline{\text{by low-mass stars (1 < M < 5 M}_\odot)}$

Normalization of the calculated ^{15}N abundance to the solar value requires an ejection rate of ^{15}N corresponding to M_{ej} (^{15}N)/M_{ej} (1 < M < 5 M$_\odot$) = 3.4 x 10^{-5}. The results are presented in table 2. One sees that the $^{12}\text{C}^{15}\text{N}/^{13}\text{C}^{14}\text{N}$ ratio obtained in this case reproduces the present interstellar value, while for the galactic center it is about two times larger than the observed 1σ upper limit.

C - $\underline{\text{The ratio of }^{15}\text{N production is proportional to the rate of mass ejectio}}$
$\underline{\text{by high-mass stars (M > 5 M}_\odot) \text{ (models C1 and C2)}}$.

Model C1 assumes no dependence of the ^{15}N production with the initial chemical composition. Normalization requires M_{ej}(^{15}N)/M_{ej}(M > 5 M$_\odot$) = 4.2 x 10^{-5}. The results can be seen in table 2, and are in agreement with the observations. As expected, since the rate of mass ejection by massive stars is relatively small at the galactic center, the production of ^{15}N is thus much smaller than in cases A and B.

We have also made some calculations as an alternative (case C2) by assuming that the production of ^{15}N is proportional either to the abundance of ^{12}C and ^{16}O (which in our models are assumed to evolve similarly) or to $^{12}\text{C} + ^{16}\text{O} + ^{14}\text{N}$. The results are very similar in both

TABLE II

Typical results of the chemical evolution estimates. The reported computations are those made without infall, assuming ^{15}N destroyed during stellar processing (the assumption of non destruction of ^{15}N in stellar processes increases the ration $^{15}N/^{14}N$ vy only about 10%). In case C1 we assume $M_{ej}(^{15}N)/M_{ej}$ (M > 5 M$_\odot$) = 4.2 x 10^{-5}. In case C2 $M_{ej}(^{15}N)/M_{ej}(^{12}C+^{16}O)$ = 2 x 10^{-2} ; conclusions are not affected by making the rate of ^{15}N production proportional to that of all CNO. The calculated abundances of ^{12}C, ^{13}C and ^{14}N are those given in table 1.

	σ (m$_{gas}$/m$_{total}$)	Observed $^{12}C^{15}N/^{13}C^{14}N$	Model B (low-mass stars) ^{15}N (per mass) 10^{-6}	Model B $\frac{^{12}C^{15}N}{^{13}C^{14}N}$	Model C1 (high-mass stars) ^{15}N (per mass) 10^{-6}	Model C1 $\frac{^{12}C^{15}N}{^{13}C^{14}N}$	Model C2 (high-mass stars) ^{15}N (per mass) 10^{-6}	Model C2 $\frac{^{12}C^{15}N}{^{13}C^{14}N}$
Solar system (t=8.4 x 10^{9}yrs)	0.23	0.33	3.7	0.29	4.2	0.33	3.8	0.31
Interstellar matter in solar neighborhood (t= 13 x 10^{9}yrs)	0.11	0.19 (average)	6.6	0.13	5.9	0.12	8.2	0.18
Interstellar matter in galactic center (t=13 x 10^{9}yrs)	8 x 10^{-4}	0.026	30.	0.053	3.0	0.005	4.2	0.0079

cases and do not differ much from those obtained in the previous case.
The result obtained when ^{15}N is assumed to produced proportionally to
^{12}C $+^{16}$O is given in table 2.

D - Models with infall

For all the above cases, calculations have also been performed
using models where the effect of possible infall of external matter devoided of heavy elements is taken into account. Infall adds gas continuously, so the rate of star formation has to be higher to destroy
the excess gas (see Paper II) and the astration parameter ν has to be
increased accordingly. This higher rate of star formation has minor
effects in the solar neighborhood. But in the galactic center the
abundances of all the secondary elements (^{13}C, ^{14}N, ^{15}N) decrease
with time in roughly the same fashion while the abundance of ^{12}C remains
constant. The net effect of infall is then to increase the ^{12}C^{15}N/^{13}C^{14}N
ratio in the galactic center. A relatively large rate of infall
$\delta \simeq 10^{-11}$ M$_\odot$/(M$_\odot$ x 1 year) increases the calculated ^{15}N^{12}C/^{14}N^{13}C ratio by
a factor about 5, and is unacceptable in all cases. A smaller infall
rate $\delta = 10^{-12}$ M$_\odot$/(M$_\odot$ x yr) is not excluded by the present observations.

VI. ASTROPHYSICAL IMPLICATIONS

The previous schemes for ^{15}N production are compatible with formation
in novae or in supernovae, as we shall see now. The mass of the nova
progenitors is not well known but might be high. The mass of supernovae
progenitors is usually considered at being larger than 5 M$_\odot$. Space
distribution of novae and supernovae might tell us something about the
lifetime and mass of their progenitors, when compared to the predictions
of galactic evolution models. We have shown in Paper II that the space
distribution of supernovae is indeed compatible with their progenitors
having M > 5 M$_\odot$. If the space distribution of novae is very poorly
known in our Galaxy, it can be derived from the statistics of novae
observed in the Andromeda nebula M31. We have used for this purpose
the lists of novae published for this galaxy by Hubble (1929), Arp (1956)
and Rosino (1964) combined with the isophotometry of de Vaucouleurs (1958)

to estimate the variation of the rate of novae per unit of surface brightness as a function of radius. This distribution is presented in fig. 1 for all novae and for "fast" novae (decay faster than 0.3 mg/day) which might have a different nucleosynthetic behaviour.

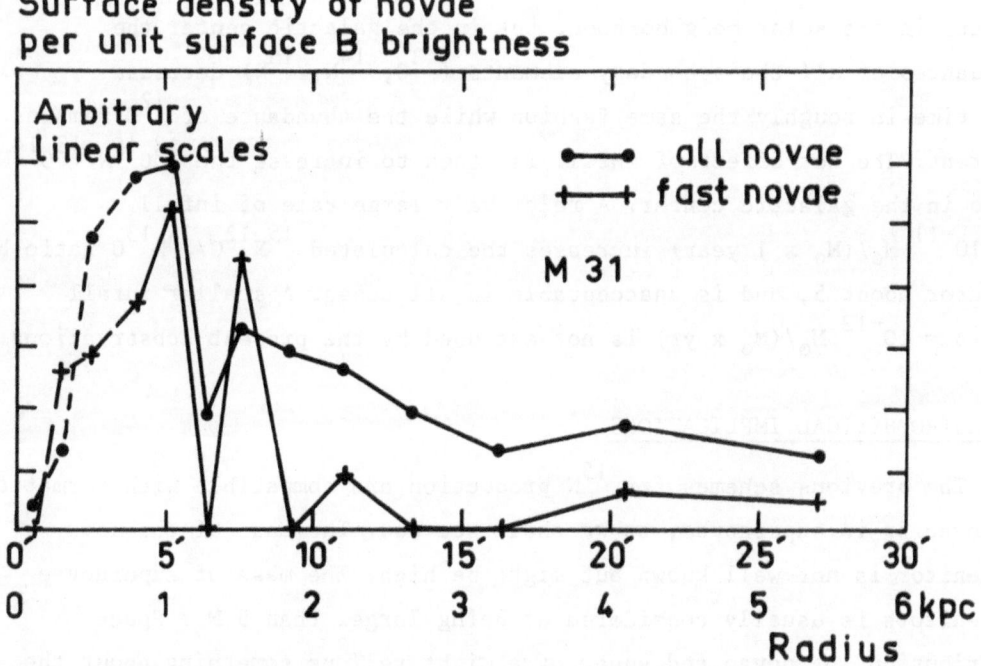

Figure 1 - Distribution of novae as a function of radius in M 31.

One sees that this number is not a strong function of radius[*]
Since the mass/luminosity ratio is about constant in the inner parts of
M 31 (see eg. Emerson, 1976) the number of novae is thus roughly proportional to the total mass. In Paper II we have shown that the rate of
mass ejection by low mass stars (< 5 $M_\odot$) per unit total mass is roughly
the same in the solar neighborhood and in the galactic center at the
present time and thus also the rate of death of stars of these masses.
(In contrast, the rate of high mass stars per unit total mass is smaller
by a factor $\sim$ 10 in the galactic center than in the solar neighborhood.)
Consequently the available data on M 31 (presumably similar to our
Galaxy in this respect) are roughly consistent with a rate of novae
being proportional to the rate of mass ejection from (or death of)
stars with 1 < M < 5 $M_\odot$.

Cases A and B of the previous section are thus both consistent
with ^{15}N being produced in novae, on the basis of their space distribution (we have made calculations for these two cases, in our ignorance
of the evolutionary properties of novae). Case A (rate independent of
time) is clearly excluded, but case B is only marginally inconsistent
with the ^{15}N observations. This case implies that the mass of ^{15}N produced per nova explosion is 2 x 10^{-7} M_O, a fraction 2 x 10^{-3} of the total
ejected mass which is of the order of 10^{-4} $M_\odot$ in the most recent models
of Starrfield et al, (to be published). This is quite compatible with
what is obtained from nucleosynthesis predictions of these models.

The abundance of ^{15}N in the galactic center would decrease if
the progenitors of novae had larger masses. To illustrate this effect,
one may assume eg. that the rate of novae is proportional to the rate
of supernovae (case C1, rate of novae/rate of supernovae $\approx$ 1.3 x 10^{3}) :
the ^{15}N abundance is lowered by a factor 10 in the galactic center

---

[*] The decrease of the number of novae for r<3' is probably due to a
selection effect (difficulty of detecting novae against the high background of the central regions).

relative to case B. However the novae mass range does not account
now for the distribution of novae observed in M 31. An intermediate
mass spectrum for novae progenitors (say 3 - 8 $M_\odot$) would be a possible
compromise. Another possibility is that only a small fraction of novae
characterized by large-mass progenitors would be active in ^{15}N production :
each of them would release perhaps 10 times more ^{15}N than in the case
B above. This is not excluded by current models of nucleosynthesis of
novae. Indeed the production of ^{15}N depends strongly on the temperature-
density profile achieved in the outbursts : in cases where the maximum
temperature is too low (T < 1.5 x 10^8K) there is no significant production
of ^{15}N.

Alternative astrophysical sites for ^{15}N production are supernovae,
where ^{15}N may be produced by explosive hydrogen and helium burning. Case
C2 corresponds to this possibility, and gives results consistent with
the observed upper limit of $^{12}C^{15}N/^{13}C^{14}N$ in the galactic center. In
this case, each supernova explosion would release about 10^{-4} $M_\odot$ of
^{15}N. To achieve such a production not only the temperature has to be
larger than 10^8K but the density has to be at least of the order of
1 g cm^{-3} (Audouze, et al, 1973). This is rather large compared to the
densities reached normally in the hydrogen-rich zones ; the only way
to produce ^{15}N in supernovae is to make it in explosive helium-burning
zones which should have higher densities following the processes investigated by Howard et al,(1971) and Arnould and Beelen (1974). More realistic
models of supernova nucleosynthesis have to be made before one can argue
more in favour of this hypothesis.

In conclusion, the observations of HCN isotopic ratios, although
they show that the $^{15}N/^{14}N$ ratio changes much more than $^{13}C/^{12}C$ from
the solar neighborhood to the galactic center, cannot be interpreted in
a definitive way. Our analysis seems to exclude a rate of infall larger
than 10^{-12} $M_\odot/(M_\odot$ x year). Both possible scenarios for producing N raise
problems. The apparent proportionality to the total mass of the rate
of novae in M 31 (and thus probably the Galaxy) is difficult to reconci-
le with the observed decrease of the $^{12}C^{15}N/^{13}C^{14}N$ ratio in the galactic

center; on the other hand ^{15}N nucleosynthesis in supernovae would improve the fit with observations, but is not well understood. An improvement of the upper limit of the $^{12}C^{15}N/^{13}C^{14}N$ ratio in the galactic center should clarify the situation : if the actual value would turn out to be significantly lower than the present upper limit, ^{15}N production in novae would be ruled out. But the present upper limit does not allow to eliminate the novae as the possible main producers of ^{15}N.

We thank the group of the Bell Lab. for sending us their paper prior to publication. We are indebted to Drs. Sumner Starrfield and James W. Truran for enlightening discussions.

REFERENCES

Arnould, M. and Beelen, W., 1974, Astron. and Astrophys. 33, 215

Arp, H.C., 1956, Astron. J., 61, 15

Audouze, J. and Fricke, K.J., Ap. J., 186, 239

Audouze, J., Lequeux, J. and Vigroux, L., 1975, Astron. and Astrophys., 43, 71 (Paper I)

Audouze, J., Truran, J.W. and Zimmerman, B.A., 1973, Ap. J., 184, 493

Clark, F.O, Buhl, D. and Snyder, L.E., 1974, Ap. J., 190, 545

Emerson, D.T., 1976, M.N.R.A.S., 176, 321

Howard, W.M., Arnett, W.D., and Clayton, D.D., 1971, Ap. J., 165, 495

Hubble, E.P., 1929, Ap. J., 69, 103

Linke, R.A., Golsmith, P.F., Wannier, P.G., Wilson, R.F. and Penzias, A.A., 1976, submitted to Ap. J.

Prialnik, D., Shara, M.M. and Shaviv, G., 1976, preprint

Rosini, L., 1964, Ann. Astroph., 27, 498

Sneden, C., and Lambert, D.L., 1975, M.N.R.A.S., 170, 533

Starrfield, S., Sparks, W.M., and Truran, J.W., 1974, Ap. J. Suppl., 28, 247

Starrfield, S., Truran, J.W., Sparks, W.M., and Kutter, G.S., 1972, Ap. J., 176, 169

Talbot, R.J. and Arnett, W.D., 1973, Ap. J., 186, 69

Talbot, R.J. and Arnett, W.D., 1974, Ap. J., 190, 605

Vaucouleurs, G. de, 1958, Ap. J., 128, 465

Vigroux, L., Audouze, J., and Lequeux, J., 1976, Astron. and Astrophys., 52, 1 (Paper II)

Wannier, P.G., Penzias, A.A., Linke, R.A., Wilson, R.W., 1976, Ap. J., 204, 26.

Wilson, R.W., Penzias, A.A., Jefferts, K.B, Thaddeus, P., Kutner, M.L., 1972. Ap. J. Letters, 176, L77.

Wollman, E.R., 1973, Ap. J. 184, 773.

PART VI

CONCLUSION

CONCLUDING REMARKS

Beatrice M. Tinsley
Yale University Observatory

This has been such a long and varied session that a full summary would be impossible. Let me briefly describe some personal impressions as to what we have learned here about CNO, and what CNO isotopes can teach us about astrophysics in general.

Two contrasting features have characterized this meeting. On the one hand, there has been ample opportunity for controversies on specific issues, ranging from nuclear reaction rates to the influence of novae on galactic evolution, to be argued in detail among the experts. And on the other hand, the diversity of fields represented here has given each of us a new appreciation of some of the difficulties and uncertainties in the theory and data that impinge from other specialties on our own work. The critical insights will prove useful in contexts well beyond the study of CNO isotopes.

In these remarks, I shall emphasize the interdisciplinary aspect of the information presented here.

We have learned that a striking variety of processes may affect a reported isotopic or elemental abundance ratio involving CNO. I use the adjective "reported" because the major problem is often to find out what the true abundance ratios are, before one can even ask about their origins. The following effects have been discussed, in particular:

Nucleosynthesis.

Stellar evolution, including mixing, mass loss, and interactions in binary systems.

Geochemistry.

Interstellar physics, including the physical state of clouds and other regions of the interstellar medium , and the extent and persistence of abundance inhomogeneities.

Jean Audouze (ed.), CNO Isotopes in Astrophysics, 175-180. All Rights Reserved.
Copyright © 1977 by D. Reidel Publishing Company, Dordrecht-Holland.

Radiative transfer.

Selection effects.

Interstellar chemistry, including the formation of molecules and grains, and isotopic fractionation during these processes.

Galactic evolution, including the effects of stellar nucleosynthesis and mass loss, and possible "infall" of matter into the Galaxy.

Almost every chapter of astrophysics is invoked somewhere here (except, surprisingly, cosmology - and it takes little imagination to introduce that ubiquitous subject: for example, the age of the Universe has a lower bound given by the ages of globular clusters, which could be severly affected if unforseen amounts of stellar mixing are responsible for some of the surface abundance anomalies discussed!).

The chief problem for understanding many apparent abundance ratios is clearly to determine <u>which</u> of the above categories provides the dominant influence. A few examples will illustrate the necessity to resolve important ambiguities of this type; by discussing only a few points here, I do not mean to deny importance to the many other aspects of this problem raised in the papers contained in these proceedings - I hope only to indicate how one might bring more order to the considerable diversity of viewpoints.

1. Several speakers, including Demarque, Kraft, and Dearborn, have discussed anomalous elemental and isotopic <u>abundance ratios on stellar surfaces</u>. Here it is crucial to determine whether such ratios reflect initial interstellar abundance variations, or mixing out of products of nucleosynthesis in the star itself. The former alternative, if applicable to the globular cluster stars in particular, would imply remarkable abundance inhomogeneities in the pre-stellar gas.

The latter explanation leads to key estimates of the contributions of various stars to enrichment of the interstellar medium, and later generations of stars, during the course of galactic evolution. These estimates are unfortunately very sensitive not only to the surface abundances, but also to the rate of stellar mixing, the stages of interior evolution at which it occurs, and the stellar mass loss rates at those times. For example, there is evidence that low-mass carbon stars have mixed core ^{12}C from the triple-alpha process into their envelopes (Kilston, 1975); with even a tiny amount of such mixing, these stars could be a major source of ^{12}C in the Galaxy, or a source of ^{14}N (then a primary product!) if the material is processed toward CNO equilibrium before most of it is shed by the star. In general, in order to use surface abundances to make empirical estimates of interstellar enrichment rates, the abundances must be supplemented by

values of the mass of envelope material characterized by that
abundance (the initial mass of the star and the amount of mass lost
already), and the fraction that is lost without further nuclear
processing. Lithium provides an interesting example from outside the
field of CNO: this element is very fragile against destruction in the
hot depths of stellar envelopes, but nevertheless the "super-Li" giant
stars could be a significant source of ^{7}Li if they are losing mass
at plausible rates during the time of lithium enhancement at their
surfaces (Scalo, 1976). The papers here on novae gave further examples.

2. R. N. Clayton has reviewed the question of whether some
oxygen isotopic anomalies in meteorites reflect admixtures of material
with pure ^{16}O that survived from the time of nucleosynthesis of ^{16}O,
or reflect very efficient isotopic fractionation within the solar
system. The former, preferred, explanation means that matter can
sometimes avoid being homogenized in the interstellar medium during
its journey from a site of nucleosynthesis to incorporation into
another star; one is reminded of the possibility of the formation in
supernova envelopes of grains that are refractory enough to survive
into the solar system (cf. D. D. Clayton, 1975). Either explanation
of the oxygen anomalies casts a disturbing measure of doubt on the
"universality" of many accepted isotopic abundances that are derived
from meteorites. We are warned not to be too confident in meteoritic
material as a sample of exactly the average interstellar abundance
distribution 4.6 billion years ago!

There is, of course, strong evidence for significant variations
in composition of the interstellar medium on stellar-mass scales: the
scatter of metallicities in stars of a given age is an obvious example.
By overlooking substantial inhomogeneities, could we be trying too hard
to gear theories of nucleosynthesis to match the solar-system abundance
distribution (hopefully called "cosmic"), and theories of galactic
evolution to explain detailed abundance differences between the sun
and the present interstellar medium? Truran has discussed here the
possibility that certain isotopic abundances in the solar system are
due to a supernova that occurred nearby just before the system
condensed. I would not be surprised if other abundances should be
considered in a similar vein, for the following particular reasons in
addition to the general nature of inhomogeneities: There is evidence
that the sun has an above-average metal abundance for a star of its
age (e.g. Mayor, 1976; Fig. 1), which would imply that the solar
system contains unusually great proportions of elements liberated in
a supernova explosion, whose products were not fully mixed to give
"average" interstellar abundances before some of the polluted material
was isolated in the condensing system. Furthermore, with very few
exceptions (e.g. van den Bout's and Shuter's papers here), there are
almost no claims for a ratio of ^{12}C/^{13}C as high as 90 anywhere else.
A tentative and unorthodox suggestion is therefore that the solar
system has an overabundance of ^{12}C.

3. An important problem is whether the interstellar abundance

ratio $^{12}C/^{13}C \sim 40$, found in many molecules (as discussed in several
papers, e.g. the review by Wannier) reflects the true isotopic ratio
in average present interstellar material, or is significantly too low
because of chemical fractionation. The latter possibility was argued
persuasively here by Watson.

If we accept the former possibility, and the solar system ratio
of 90 as typical of material in this part of the Galaxy 4.6 billion
years ago, then models for galactic evolution can account very nicely
for the implied relatively faster enrichment in ^{13}C, as reported by
Lequeux. These models, especially when provided with further
constraints based on other elements, lead to some very interesting
conclusions concerning evolutionary processes in the Galaxy. Yet
before they are accepted, we should seriously consider whether the
true interstellar ratio of $^{12}C/^{13}C$ is still nearer to 90.

The potential importance of chemical fractionation leads to a
point often made by Fowler: we should reserve the term "chemical
evolution" for truly chemical processes, such as fractionation and
molecule formation, and refer to changes in average abundances over
the lifetime of the Galaxy in another way. The term "nuclear evolution"
has been suggested, but needs improvement because of its possible
confusion with studies of galactic nuclei, and its neglect of processes
such as star formation and gas flows, which are as important as
nucleosynthesis in determining the evolution of abundances.

The critical question here is, which kind of evolution is
responsible for the difference between the $^{12}C/^{13}C$ abundance ratios
in the solar system and in interstellar molecules? There are plausible
explanations in terms of either chemical or "nuclear" evolution, both
beset by the possibility that 90 is an atypically great value for
4.6 billion years ago; maybe all three effects - fractionation,
nucleosynthesis and ejection by stars, and inhomogeneities - are
significant.

These ambiguities with carbon isotopes recall a set of parallel
problems with other elements - whose interpretation is generally
roughly antiparallel to the common interpretation of $^{12}C/^{13}C$ in terms
of "nuclear" evolution! I refer to the fact that almost all discussions
of interstellar abundances of heavier elements assume that the gas-
phase abundances are below solar entirely because of grain formation.
By analogy with the alternatives for carbon, it would be worth
considering, in addition to truly chemical processes, possible effects
of galactic "nuclear" evolution and the presence of some anomalies in
the solar system abundances used as the "cosmic" standards. As an
example of the latter consideration, Talbot and Arnett (1973)
suggested that part of the apparent depletion could be due to
excessive abundances of heavy elements in the sun and other stars; as
another example, the evidence for excessive metallicity in the sun
(noted above) may be relevant. On the evolutionary side, there is
the elusive possibility of a change in the average interstellar metal

abundance within the past 4.6 billion years (McClure and Tinsley 1976), or of changes in relative abundances of various elements. Just as evolutionary models can plausibly (but not uniquely) predict large increases in the abundances of CNO isotopes whose major source is in low-mass stars (Audouze et al., 1975; Vigroux et al., 1976), so should one look for important changes in other elements that may be synthesized by stars below about $2M_\odot$. The lifetime of the synthesizing stars is generally predicted to be a more important factor in relative evolution of various abundances, than whether the elements concerned are primary, secondary, etc. (cf. Tinsley, 1976). Yet another serious ambiguity in the interpretation of apparent depletion factors was discussed here by Steigman, who noted that with certain considerations of the curve-of-growth analysis, the interstellar abundances could be normal after all!

I have made this digression into the question of apparent depletion into grains chiefly in order to illustrate again how studies of CNO isotopes can provide insights in a broader context. In particular, the discussions of CNO have raised a formidable list of factors to be weighed before any explanation is adopted for the apparent departure of an abundance ratio from its solar-system value.

The tone of this conference has been to turn up complications and problems even in situations where satisfactory data and theory seemed to exist - resulting in a powerful stimulus to future research. One astronomer, leaving the meeting as it ended at 6 p.m., was overheard to say, "This has been one of those sessions I've left knowing less than when I came in"- surely a compliment to the organizer and to the participants, who gave us some of the wisdom of knowing that we know (almost) nothing.

All who were present are very grateful to everyone who made possible such a richly varied session: to the speakers for a series of excellent talks; to members of the audience for invaluable comments; to Willy Fowler, whose own research and inspiration lay behind many of the contributions, for his stimulating chairmanship; and to Jean Audouze, for having the insight to plan such a successfully diverse program. My personal thanks go to Jean for the invitation to make closing comments, to many astronomers in Grenoble and elsewhere for discussions of their various fields of expertise, and to those speakers (too numerous to refer to above) who gave me the benefit of advance copies of their papers; I acknowledge financial support from the National Science Foundation (Grant AST75-16329) and the Alfred P. Sloan Foundation.

REFERENCES

Audouze, J., Lequeux, J., and Vigroux, L.: 1975, Astron. Astrophys. 43,71.
Clayton, D. D.: 1975, Astrophys. J. 199, 765.
Kilston, S.: 1975, Publ. Astron. Soc. Pacific 87, 189.

Mayor, M.: 1976, Astron. Astrophys. 48, 301.
McClure, R. D., Tinsley, B. M.: 1976, Astrophys. J. 208, 480.
Scalo, J. M.: 1976, Astrophys. J. 206, 795.
Talbot, R. J., Arnett, W. D.: 1973, Astrophys. J. 186, 69.
Tinsley, B. M.: 1976, Astrophys. J. 208, 797.
Vigroux, L., Audouze, J., and Lequeux, J.: 1976, Astron. Astrophys.
      (in press).

INDEX OF SUBJECTS

INDEX OF NAMES

The page number of the beginning of each chapter is underlined.

ASTROPHYSICS AND SPACE SCIENCE LIBRARY

Edited by

J. E. Blamont, R. L. F. Boyd, L. Goldberg, C. de Jager, Z. Kopal, G. H. Ludwig, R. Lüst,
B. M. McCormac, H. E. Newell, L. I. Sedov, Z. Švestka, and W. de Graaff

24. B. M. McCormac (ed.), *The Radiating Atmosphere. Proceedings of a Symposium Organized by the Summer Advanced Study Institute, held at Queen's University, Kingston, Ontario, August 3–14, 1970.* 1971, XI + 455 pp.

25. G. Fiocco (ed.), *Mesospheric Models and Related Experiments. Proceedings of the 4th ESRIN-ESLAB Symposium, held at Frascati, Italy, July 6–10, 1970.* 1971, VIII + 298 pp.

26. I. Atanasijević, *Selected Exercises in Galactic Astronomy.* 1971, XII + 144 pp.

27. C. J. Macris (ed.), *Physics of the Solar Corona. Proceedings of the NATO Advanced Study Institute on Physics of the Solar Corona, held at Cavouri-Vouliagmeni, Athens, Greece, 6–17 September 1970.* 1971, XII + 345 pp.

28. F. Delobeau, *The Environment of the Earth.* 1971, IX + 113 pp.

29. E. R. Dyer (general ed.), *Solar-Terrestrial Physics/1970. Proceedings of the International Symposium on Solar-Terrestrial Physics, held in Leningrad, U.S.S.R., 12–19 May 1970.* 1972, VIII + 938 pp.

30. V. Manno and J. Ring (eds.), *Infrared Detection Techniques for Space Research. Proceedings of the 5th ESLAB-ESRIN Symposium, held in Noordwijk, The Netherlands, June 8–11, 1971.* 1972, XII + 344 pp.

31. M. Lecar (ed.), *Gravitational N-Body Problem. Proceedings of IAU Colloquium No. 10, held in Cambridge, England, August 12–15, 1970.* 1972, XI + 441 pp.

32. B. M. McCormac (ed.), *Earth's Magnetospheric Processes. Proceedings of a Symposium Organized by the Summer Advanced Study Institute and Ninth ESRO Summer School, held in Cortina, Italy, August 30–September 10, 1971.* 1972, VIII + 417 pp.

33. Antonin Rükl, *Maps of Lunar Hemispheres.* 1972, V + 24 pp.

34. V. Kourganoff, *Introduction to the Physics of Stellar Interiors.* 1973, XI + 115 pp.

35. B. M. McCormac (ed.), *Physics and Chemistry of Upper Atmospheres. Proceedings of a Symposium Organized by the Summer Advanced Study Institute, held at the University of Orléans, France, July 31–August 11, 1972.* 1973, VIII + 389 pp.

36. J. D. Fernie (ed.), *Variable Stars in Globular Clusters and in Related Systems. Proceedings of the IAU Colloquium No. 21, held at the University of Toronto, Toronto, Canada, August 29–31, 1972.* 1973, IX + 234 pp.

37. R. J. L. Grard (ed.), *Photon and Particle Interaction with Surfaces in Space. Proceedings of the 6th ESLAB Symposium, held at Noordwijk, The Netherlands, 26–29 September, 1972.* 1973, XV + 577 pp.

38. Werner Israel (ed.), *Relativity, Astrophysics and Cosmology. Proceedings of the Summer School, held 14–26 August, 1972, at the BANFF Centre, BANFF, Alberta, Canada.* 1973, IX + 323 pp.

39. B. D. Tapley and V. Szebehely (eds.), *Recent Advances in Dynamical Astronomy. Proceedings of the NATO Advanced Study Institute in Dynamical Astronomy, held in Cortina d'Ampezzo, Italy, August 9–12, 1972.* 1973, XIII + 468 pp.

40. A. G. W. Cameron (ed.), *Cosmochemistry. Proceedings of the Symposium on Cosmochemistry, held at the Smithsonian Astrophysical Observatory, Cambridge, Mass., August 14–16, 1972.* 1973, X + 173 pp.

41. M. Golay, *Introduction to Astronomical Photometry.* 1974, IX + 364 pp.

42. D. E. Page (ed.), *Correlated Interplanetary and Magnetospheric Observations. Proceedings of the 7th ESLAB Symposium, held at Saulgau, W. Germany, 22–25 May, 1973.* 1974, XIV + 662 pp.

43. Riccardo Giacconi and Herbert Gursky (eds.), *X-Ray Astronomy.* 1974, X + 450 pp.

44. B. M. McCormac (ed.), *Magnetospheric Physics. Proceedings of the Advanced Summer Institute, held in Sheffield, U.K., August 1973.* 1974, VII + 399 pp.

45. C. B. Cosmovici (ed.), *Supernovae and Supernova Remnants. Proceedings of the International Conference on Supernovae, held in Lecce, Italy, May 7–11, 1973.* 1974, XVII + 387 pp.

46. A. P. Mitra, *Ionospheric Effects of Solar Flares.* 1974, XI + 294 pp.

47. S.-I. Akasofu, *Physics of Magnetospheric Substorms.* 1977, XVIII + 599 pp.

48. H. Gursky and R. Ruffini (eds.), *Neutron Stars, Black Holes and Binary X-Ray Sources.* 1975, XII + 441 pp.

49. Z. Švestka and P. Simon (eds.), *Catalog of Solar Particle Events 1955–1969. Prepared under the Auspices of Working Group 2 of the Inter-Union Commission on Solar-Terrestrial Physics.* 1975, IX + 428 pp.

50. Zdeněk Kopal and Robert W. Carder, *Mapping of the Moon.* 1974, VIII + 237 pp.

51. B. M. McCormac (ed.), *Atmospheres of Earth and the Planets. Proceedings of the Summer Advanced Study Institute, held at the University of Liège, Belgium, July 29–August 8, 1974.* 1975, VII + 454 pp.

52. V. Formisano (ed.), *The Magnetospheres of the Earth and Jupiter. Proceedings of the Neil Brice Memorial Symposium, held in Frascati, May 28–June 1, 1974.* 1975, XI + 485 pp.

53. R. Grant Athay, *The Solar Chromosphere and Corona: Quiet Sun*. 1976, XI + 504 pp.
54. C. de Jager and H. Nieuwenhuijzen (eds.), *Image Processing Techniques in Astronomy. Proceedings of a Conference, held in Utrecht on March 25–27, 1975*, XI + 418 pp.
55. N. C. Wickramasinghe and D. J. Morgan (eds.), *Solid State Astrophysics. Proceedings of a Symposium, held at the University College, Cardiff, Wales, 9–12 July 1974*. 1976, XII + 314 pp.
56. John Meaburn, *Detection and Spectrometry of Faint Light*. 1976, IX + 270 pp.
57. K. Knott and B. Battrick (eds.), *The Scientific Satellite Programme during the International Magnetospheric Study. Proceedings of the 10th ESLAB Symposium, held at Vienna, Austria, 10–13 June 1975*. 1976, XV + 464 pp.
58. B. M. McCormac (ed.), *Magnetospheric Particles and Fields. Proceedings of the Summer Advanced Study School, held in Graz, Austria, August 4–15, 1975*. 1976, VII + 331 pp.
59. B. S. P. Shen and M. Merker (eds.), *Spallation Nuclear Reactions and Their Applications*. 1976, VIII + 235 pp.
60. Walter S. Fitch (ed.), *Multiple Periodic Variable Stars. Proceedings of the International Astronomical Union Colloquium No. 29, Held at Budapest, Hungary, 1–5 September 1975*. 1976, XIV + 348 pp.
61. J. J. Burger, A. Pedersen, and B. Battrick (eds.), *Atmospheric Physics from Spacelab. Proceedings of the 11th ESLAB Symposium, Organized by the Space Science Department of the European Space Agency, held at Frascati, Italy, 11–14 May 1976*. 1976, XX + 409 pp.
62. J. Derral Mulholland (ed.), *Scientific Applications of Lunar Laser Ranging. Proceedings of a Symposium held in Austin, Tex., U.S.A., 8–10 June, 1976*. 1977, XVII + 302 pp.